ECONOPHYSICS OF INCOME AND WEALTH DISTRIBUTIONS

The distribution of wealth and income is never uniform, and philosophers and economists have tried for years to understand the reasons and formulate remedies for such inequalities. This book introduces the elegant and intriguing kinetic exchange models that physicists have developed to tackle these issues.

This is the first monograph in econophysics focused on the analyses and modelling of these distributions, and is ideal for physicists and economists. It explores the origin of economic inequality. It is written in simple, lucid language, with plenty of illustrations and in-depth analyses, making it suitable for researchers new to this field as well as more specialized readers.

BIKAS K. CHAKRABARTI is a Senior Professor of Theoretical Condensed Matter Physics at the Saha Institute of Nuclear Physics, and a visiting Professor of Economics at the Indian Statistical Institute. He has research interests in statistical physics, condensed matter physics, computational physics and econophysics.

ANIRBAN CHAKRABORTI is an Associate Professor at the Quantitative Finance Group, École Centrale Paris. He has research interests in statistical physics, econophysics and quantum physics.

SATYA R. CHAKRAVARTY is a Professor in the Economic Research Unit of the Indian Statistical Institute. His main areas of research interests are welfare economics, public economics, mathematical finance, industrial organization and game theory.

ARNAB CHATTERJEE is a Postdoctoral Researcher at Aalto University. He has research interests in statistical physics, and its application to condensed matter and social sciences.

ECONOPHYSICS OF INCOME AND WEALTH DISTRIBUTIONS

BIKAS K. CHAKRABARTI
Saha Institute of Nuclear Physics

ANIRBAN CHAKRABORTI
École Centrale Paris

SATYA R. CHAKRAVARTY
Indian Statistical Institute

ARNAB CHATTERJEE
Aalto University

CAMBRIDGE
UNIVERSITY PRESS

CAMBRIDGE
UNIVERSITY PRESS

Shaftesbury Road, Cambridge CB2 8EA, United Kingdom

One Liberty Plaza, 20th Floor, New York, NY 10006, USA

477 Williamstown Road, Port Melbourne, VIC 3207, Australia

314–321, 3rd Floor, Plot 3, Splendor Forum, Jasola District Centre, New Delhi – 110025, India

103 Penang Road, #05–06/07, Visioncrest Commercial, Singapore 238467

Cambridge University Press is part of Cambridge University Press & Assessment, a department of the University of Cambridge.

We share the University's mission to contribute to society through the pursuit of education, learning and research at the highest international levels of excellence.

www.cambridge.org
Information on this title: www.cambridge.org/9781107013445

First published 2013

A catalogue record for this publication is available from the British Library

Library of Congress Cataloging-in-Publication data
Chakrabarti, B. K. (Bikas K.), 1952–
Econophysics of income and wealth distributions / Bikas K. Chakrabarti, Anirban Chakraborti, Satya R. Chakravarty, Arnab Chatterjee.
pages cm
Includes bibliographical references and index.
ISBN 978-1-107-01344-5
1. Income distribution. 2. Wealth. 3. Econophysics.
I. Chakraborti, Anirban. II. Chakravarty, Satya R. III. Chatterjee, Arnab. IV. Title.
HB523.C43 2013
339.2 – dc23 2012032861

ISBN 978-1-107-01344-5 Hardback

Contents

Preface

An imbalance between rich and poor is the oldest and most fatal ailment of all republics.

Plutarch, ancient Greek biographer (c. 46–120 CE)

Why does this imbalance exist in the first place? Why are a few rich and many poor? For centuries we have borne the effects of this inequality. We know neither the cause nor the solution to this elusive problem. From philosophers to economists, many have vehemently tried for ages to understand the reasons and formulate remedies for such inequalities. No doubt, great efforts have been made to tackle this multifaceted problem, but the situation has been analogous to fighting the Greek mythological monster Hydra, who grows two heads in place of an injured one. Overcoming this problem, indeed, seems to be a Herculean task!

Heraclitus[1] said, 'change is the only constant'. Putting our faith in him, one might have expected things to change drastically, and the inequality to even disappear at some point in time! Strangely, this has not been the case. We find that inequality has been a universal and robust phenomenon – not bound by either time or geography. Fortunately for scholars, it has a few statistical regularities, most of which have been recorded in the past 115 years or so. Owing to the seminal works of Pareto (1897) and Gibrat (1931), one can now identify certain regularities in the income and wealth distributions over a wide range of societies and time periods. Physicists have come up with some very elegant and intriguing kinetic exchange models in recent times to shed some light on these observations. Our intention is to describe these developments in this book.

Standard economic theory would like to consider that the activities of individual agents are driven by the utility maximization principle. The alternative picture proposed by physicists is that the agents can be simply viewed as gas particles exchanging 'money', in the place of energy, and trades as money (energy)

[1] Ancient Greek philosopher (*c.* 535–475 BCE).

conserving two-body scatterings, as in the entropy maximization-based kinetic theory of gases. This qualitative analogy seems to be quite old, and both economists and natural scientists had already noted it earlier in various contexts. However, this equivalence between the two maximization principles has gained firmer ground only recently.

When tested with empirical data from various countries, just pure kinetic exchange models fall short of accommodating the Pareto tail. However, the introduction of ‘saving propensity’ (in various forms) in such kinetic exchange models enables one to successfully explain several of the observed features, including the much desired Pareto tail. A direct link between the saving propensity distribution and the inequality can also be established. The subsequent developments in the analysis of these models further established many intriguing features in the observed data. The mathematical structures of these models and their economic implications are now being investigated extensively. As mentioned above, the discovery of the equivalence of the physical entropy and the utility or psychological satisfaction, and their corresponding maximization principles, marks the entry of the kinetic exchange models of market in the domain of macroeconomics.

Interestingly, the economic inequality is a natural outcome of this framework of stochastic kinetics of trading processes in the market, independent of any exogenous factors. Thus, the kinetic exchange models described in this book demonstrate how inequality may arise. They also indicate how its effects may be partially reduced by modifying the saving habits.

The book is organized as follows: the first chapter introduces the topic to the readers. In Chapter 2, a detailed presentation of the recorded data and analyses of the income and wealth distributions across various countries in different time periods is given. In Chapter 3, some of the major recent attempts to set up the physics-inspired many-body dynamical models for income or wealth exchanges, amongst the agents in the market or network, are discussed. In Chapter 4, the details of the numerical results for the kinetic exchange models for asset or income among the agents in the market are presented. Then, Chapter 5 gives the detailed analytical structure of such kinetic exchange models for the income and wealth distributions. Chapter 6 shows how, in two-person, two-commodity trading dynamics, the Cobb–Douglas utility maximization leads to the same kinetic exchange dynamics with uniform saving propensity, discussed in the earlier chapters. In Chapter 7, these kinetic exchange modelling approaches for income and wealth distributions leading to the economic inequalities are reviewed in terms of economics of income generation and development. Finally, we present an outlook with a brief summary of the chapters, a few discussions on new directions and open problems in the last chapter.

We are extremely grateful to all our collaborators: Urna Basu, Pratip Bhattacharyya, Anindya Sundar Chakrabarti, Guido Germano, Asim Ghosh,

Sanchari Goswami, Els Heinsalu, Aymen Jedidi, Kimmo Kaski, Mehdi Lallouache, Subhrangshu Sekhar Manna, Sugata Marjit, Marco Patriarca, Srutarshi Pradhan, Parongama Sen, Sitabhra Sinha and Robin Stinchcombe, for their contributions to these developments. We acknowledge John Angle, J. Barkley Rosser Jr., Banasri Basu, Arnab Das, Kishore Dash, Deepak Dhar, Mauro Gallegati, Kausik Gangopadhyay, Abhijit Kargupta, Thomas Lux, Matteo Marsili, Pradeep K. Mohanty, Peter Richmond, Dietrich Stauffer, Victor M. Yakovenko and Sudhakar Yarlagadda for their comments and criticisms on our work, from time to time. We are indebted to several scientists, especially Victor M. Yakovenko, who granted us the permission to use their works and figures liberally. We are grateful to Esteban Guevara and Gayatri Tilak for providing invaluable help during the preparation of the manuscript. We heartily thank Soumyajyoti Biswas, Anindya Sundar Chakrabarti, Marco LiCalzi and Gayatri Tilak for critical reading of the manuscript. We also express our gratitude to Anindya Sundar Chakrabarti for helping us rewrite Chapter 6. We are thankful to the Centre for Applied Mathematics and Computational Science Project of the Saha Institute of Nuclear Physics, Kolkata, for generous funding for the research activities in econophysics.

We hope that researchers, especially the younger ones, will find the ideas described in this book intriguing enough to inspire them to do further research and take up the Herculean challenge of solving this chronic problem, which is one of the pertinent sources of tragedy for human civilization.

Kolkata, India	*Bikas K. Chakrabarti*
Châtenay-Malabry, France	*Anirban Chakraborti*
Kolkata, India	*Satya R. Chakravarty*
Espoo, Finland	*Arnab Chatterjee*

1
Introduction

> Ill fares the land, to hastening ills a prey,
> Where wealth accumulates, and men decay.
>
> *Oliver Goldsmith,*
> *Anglo-Irish writer (1730–74)*

It would be difficult to find any society or country where income or wealth is equally distributed among its people. Socioeconomic inequality is not limited to modern times; it has been a persistent fact, and a constant source of irritation to most, since time immemorial.

The issue of inequality in terms of income and wealth is perhaps the most fiercely debated subject in economics. Economists and philosophers have spent much time on the normative aspects of this problem (Rawls 1971; Scruton 1985; Sen 1999; Foucault 2003). The direct and indirect effects of inequality on society have also been studied extensively. In particular, the effects of inequality on the growth of the economy (Benabou 1994; Aghion *et al.* 1999; Barro 1999; Forbes 2000) and on the econopolitical scenario (Blau and Blau 1982; Alesina and Rodrik 1992; Alesina and Perotti 1996; Benabou 2000) have attracted major attention. Relatively less emphasis has been put on the sources of the problem itself. There are several non-trivial issues and open questions related to this observation: How are income and wealth distributed? What are the forms of the distributions? Are they universal, or do they depend upon specific conditions of a country? Perhaps the most important question is: if inequality is universal (as some of its gross features indicate), then what is the reason for such universality?

Such questions have intrigued many personalities in the past. More than a century ago, Pareto made extensive studies in Europe and found that wealth distribution follows a power law tail for the richer sections of society (Pareto 1897), known now as the Pareto law. Separately, Gibrat (1931) worked on the same problem, and he proposed a 'law of proportionate effect'. Much later, Champernowne also considered this problem systematically and came up with a probabilistic theory to justify Pareto's claim (Champernowne 1953; Champernowne and Cowell 1998).

It was subsequently found in numerous studies that the distributions of income and wealth indeed possess some globally stable and robust features (for a review, see Yakovenko and Barkley Rosser 2009). In general, the bulk of the distribution of both income and wealth seems to fit both the log-normal and the gamma distributions reasonably well. Economists usually prefer the log-normal distribution (Gini 1921; Montroll and Shlesinger 1982), whereas statisticians (Hogg *et al.* 2007) and, more recently, physicists (Chatterjee *et al.* 2005b; Chatterjee and Chakrabarti 2007b; Yakovenko and Barkley Rosser 2009) tend to rely more on alternate forms such as the gamma distribution (for the probability density) or the Gibbs/exponential distribution (for the cumulative distribution). The upper end of the distribution, that is, the tail of the distribution, is agreed to be described well by a power law, as was found by Pareto.

These observed regularities in the income distribution may thus indicate a 'natural' law of economics. The distribution of income $P(x)$ is defined as follows: $P(x)dx$ is the probability that, in the 'equilibrium' or 'steady state' of the system,[1] a randomly chosen person would be found to have income between x and $x + dx$. Detailed empirical analyses of the income distribution so far indicate

$$P(x) \sim x^n \exp(-x/T), \quad \text{for} \quad x < x_c, \tag{1.1}$$

and

$$P(x) \sim x^{-\alpha-1}, \quad \text{for} \quad x \geq x_c, \tag{1.2}$$

where n and α are two exponents, and T denotes a scaling factor. The latter exponent α is called the Pareto exponent and its value ranges between 1 and 3 (e.g. Aoyama *et al.* 2000; Sinha 2006). A historical account of Pareto's data and that from recent sources can be found in Richmond *et al.* (2006). The crossover point x_c is extracted from the numerical fittings of the initial gamma distribution form to the eventual power law tail. One often fits the region below x_c to a log-normal form: $\log P(x) = \text{const} - (\log x)^2$. As mentioned before, although this form is often preferred by economists, the statisticians and physicists think that the gamma distribution form fits better with the data (see Salem and Mount 1974; Hogg *et al.* 2007; Yakovenko and Barkley Rosser 2009). Figure 1.1 shows the features of the cumulative income or wealth distribution.

Most of the empirical analyses, especially with recent income data, have been extensively reviewed in Chapter 2. Compared with the empirical work done on

[1] We will often be using the terms 'equilibrium' or 'steady state' interchangeably in this book; strictly speaking, for systems that are 'non-ergodic', one can only write 'steady state'.

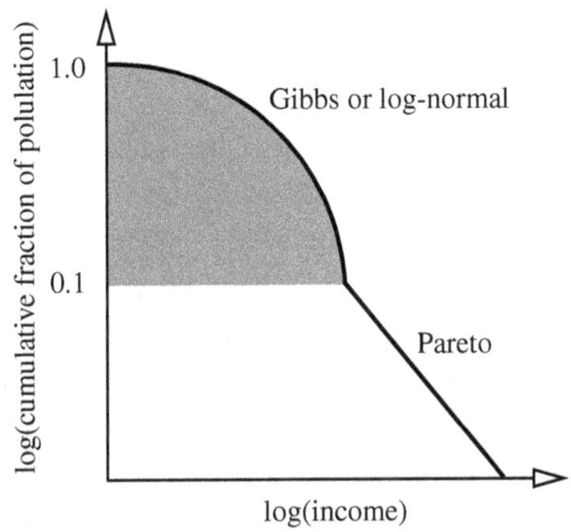

Figure 1.1 When one plots the cumulative wealth (income) distribution against the wealth (income), almost 90–95% of the population fits the Gibbs distribution, or is often fitted also to the log-normal form (Gibrat law), as indicated by the shaded region in the distribution; for the remaining (very rich) 5–10% of the population in any country, the number density falls off with their wealth (income) much more slowly, following a power law (Pareto law). It is found that approximately 40–60% of the total wealth of any economy is possessed by 5–10% of the people in the Pareto tail.

income distribution, relatively fewer studies have looked at the distribution of wealth, which consists of the net value of assets (financial holdings and/or tangible items) owned at a given instant. The lack of an easily available data source for measuring wealth, analogous to income tax returns for measuring income, means that one has to resort to indirect methods. Again, the general feature observed in the limited empirical study of wealth distribution, as presented in Chapter 2, is that of a power law behaviour for the wealthiest 5–10% of the population, and exponential or log-normal distribution for the rest of the population. The Pareto exponent, as measured from the wealth distribution, is always found to be lower than that for the income distribution, which is consistent with the general observation that, in market economies, wealth is much more unequally distributed than income (Samuelson 1998).

It is interesting to note that, when one shifts attention from the income of individuals to the income of companies, one still observes the power law tail. A study of the income distribution of Japanese firms (Aoyama *et al.* 2000; see also Aoyama *et al.* 2011) concluded that it follows a power law (with exponent value near unity, which is also often referred to as the Zipf law). Similar observation has been reported for the income distribution of companies in the USA (Axtell 2001).

These strikingly robust features of the distribution $P(x)$, in income or wealth, seem to be well established from the analyses of the enormous amount of data available today. Is it plausible that this only reflects a basic natural law, with simple physical explanation? Many econophysicists actually believe so. According to these proponents, the regular patterns observed in the income (and wealth) distribution are indeed indicative of a natural law for the statistical properties of a many-body dynamical system representing the entire set of economic interactions in a society, analogous to those previously derived for gases and liquids. By viewing the economy as a 'thermodynamic' system (Chakrabarti and Marjit 1995; Drăgulescu and Yakovenko 2000; Hayes 2002; Patriarca *et al.* 2010), one can liken income distribution to the distribution of energy among the particles in a gas. Several attempts by statisticians (e.g. Angle 1986, 2006) and economists (Bennati 1988a,b, 1993) also provide impetus to this interdisciplinary approach.

In particular, a class of kinetic exchange models (Chakraborti and Chakrabarti 2000; Chatterjee *et al.* 2003, 2004; Chakrabarti and Chatterjee 2004) have provided a simple mechanism for understanding the unequal accumulation of assets. While being simple from the perspective of economics, they have the benefit of gripping a key factor – savings – in socioeconomic interactions, which results in very different societies converging to similar forms of unequal distribution.

These simple microeconomic models, with a large number of 'agents' and the 'asset' transfer equations among the agents owing to 'trading' in such an economy, closely resemble the process of 'energy' transfer owing to 'collisions' among 'particles' like those in a thermodynamic system of ideal gas. In these models, the system is assumed to be made up of N agents with assets $\{x_i \geq 0\}$ $(i = 1, 2, \ldots, N)$. At every trade, an agent j exchanges a part Δx with another agent k chosen randomly. The total asset $X = \sum_i x_i$ is constant, as well as the average asset $\langle x \rangle = X/N$. After the exchange the new values x'_j and x'_k are $(x'_j, x'_k \geq 0)$

$$\left.\begin{aligned} x'_j &= x_j - \Delta x, \\ x'_k &= x_k + \Delta x. \end{aligned}\right\} \tag{1.3}$$

The form of the function $\Delta x = \Delta x(x_j, x_k)$ defines the underlying dynamics of the model. Figure 1.2 shows the schematic picture that captures the essence of these models.

The steady-state distribution for a system with pure random asset exchange is an exponential one, as was found by Gibbs 100 years ago (e.g. Chatterjee and Chakrabarti 2007b; Yakovenko and Barkley Rosser 2009). However, the introduction of 'saving propensity' (Chakraborti and Chakrabarti 2000) brought forth the gamma-like feature of the distribution $P(x)$ and such a random exchange model with uniform saving propensity for all agents was subsequently shown to be equivalent to a commodity clearing market in which each agent maximizes his/her own utility (Chakrabarti and Chakrabarti 2009). A further modification of the model

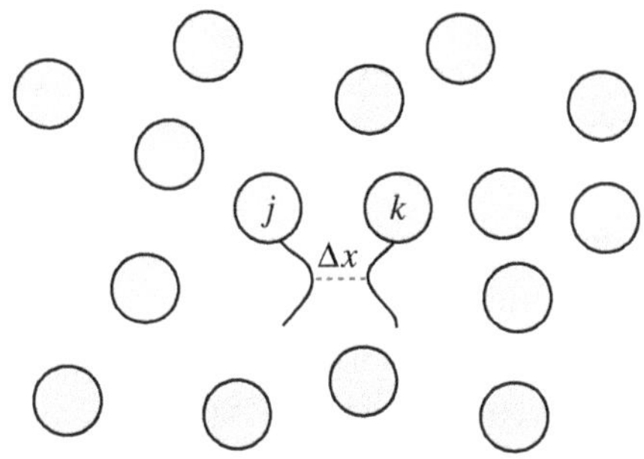

Figure 1.2 The kinetic exchange models prescribe a microscopic interaction between two units analogously to a kinetic model of gas in which, during an elastic collision, two generic particles j and k exchange an energy amount Δx, as in Eq. (1.3). Reproduced from Patriarca *et al.* (2010).

produces (Chatterjee *et al.* 2004) a power law for the upper or tail end of the distribution of money, as has been found empirically.

Several analytical aspects of this class of models have been studied (e.g. Ispolatov *et al.* 1998; Düring *et al.* 2008; Garibaldi and Scalas 2010; Lallouache *et al.* 2010b; Toscani and Brugna 2010). It is noteworthy that, at present, this is the only known class of models which, starting from the microeconomics of utility maximization and solving for the resultant dynamical equations in the line of rigorously established statistical physics, can quite reliably reproduce the major empirical features of income and wealth distributions in economies.

These developments have, of course, not gone without criticism (e.g. Hogan 2005; Lux 2005; Gallegati *et al.* 2006), and subsequent rebuttal (Richmond *et al.* 2006). In view of the embarrassing failure of mainstream economic schools to anticipate or correctly analyse the recent economic crisis, there has been some recent interest by the mainstream economic schools to revisit such physically motivated models of the market dynamics and their solutions (e.g. Lux and Westerhoff 2009).

The successive chapters of this book will review in detail the various aspects mentioned above. In Chapter 2, a detailed presentation of the recorded data and analyses of the income and wealth distributions across countries at different periods of time is given. The generic feature is, of course, as indicated in Fig. 1.1.

In Chapter 3, we discuss some of the major recent attempts to set up the physics-inspired many-body dynamical models for income or wealth exchanges among the agents in the market or network. Attempts are also made to compare the results with the established economic laws for the flow of money and the empirically observed distributions in society. In Chapter 4, we discuss in detail the numerical results for the kinetic exchange models for assets or income among the agents in the market. This development follows closely the century-old kinetic theory of gases, and models each trade as money (energy equivalent) conserving two-body collision leading to many-body steady-state or equilibrium distributions of money. As mentioned above, incorporation of saving propensity in the dynamics gives gamma-like distributions, while the dispersion in saving propensity among the agents leads to the Pareto tail from those gamma-like distributions. Chapter 5 gives the detailed analytical structure of such kinetic exchange models for income and wealth distributions. While the kinetic exchange dynamics discussed in Chapters 4 and 5 can essentially be viewed as 'entropy maximization' dynamics, it is shown to be equivalent to that following a utility maximization principle, as well. Chapter 6 shows how, in a two-person two-commodity trading dynamics, the Cobb–Douglas utility maximization leads to the same kinetic exchange dynamics with uniform saving propensity discussed in earlier chapters. These two maximization principles of physics and economics lead to identical dynamical equations. In Chapter 7, these econophysics models for income and wealth distributions leading to economic inequalities are reviewed from the perspective of economics of income generation and development. Extensive discussions on the various economic inequality indices, following the income and wealth distributions obtained in earlier chapters, are given here to cast these developments in proper economic perspectives. Finally, we present an outlook in Chapter 8, with a brief summary of the chapters and a few discussions on new directions, challenges and open problems.

2

Income and wealth distribution data for different countries

Investigations over more than a century and the recent availability of electronic databases of income and wealth distribution (ranging from a national sample survey of household assets to the income tax return data available from governmental agencies) have revealed some remarkable features. Irrespective of many differences in culture, history, social structure, indicators of relative prosperity (such as gross domestic product or infant mortality) and, to some extent, the economic policies followed in different countries, the income distribution seems to follow a particular *universal* pattern, as does the wealth distribution: after an initial rise, the number density of people rapidly decays with their income, the bulk described by a Gibbs or log-normal distribution crossing over at the very high income range (for 5–10% of the richest members of the population) to a power law, as shown in Fig. 1.1. The power law in income and wealth distribution is called the *Pareto law*, after the Italian sociologist and economist Vilfredo Pareto. The log-normal part is named as the *Gibrat law*, after the French economist Robert Gibrat. This seems to be a universal feature: from ancient Egyptian society (Abul-Magd 2002) through nineteenth-century Europe (Pareto 1897; Champernowne 1953) to modern Japan (Chatterjee *et al.* 2005b; Chakrabarti *et al.* 2006). The same is true across the globe today: from the advanced capitalist economy of USA (Chatterjee *et al.* 2005b; Kar Gupta 2006a; Richmond *et al.* 2006) to the developing economy of India (Sinha 2006).

A historical account of the empirical data analyses, followed by an account of research using recent sources, will be presented in this chapter. Country-wise studies at various time periods will be also presented. Measures of income inequality in terms of the Gini coefficient and other indices will be briefly reviewed.

2.1 What are money, wealth and income?

Let us start by considering the basic economic quantities: money, wealth and income. A common definition of *money* suggests that money is ‘a commodity accepted by general consent as a medium of economic exchange'.[1] In fact, money circulates from one economic agent (which can represent an individual, firm, country, etc.) to another, thus facilitating trade. It is 'something which all other goods or services are traded for' (for details, see Shostak 2000). Throughout history various commodities have been used as money, for these cases termed 'commodity money', which include for example rare seashells or beads and cattle (such as the cow in India). Recently, 'commodity money' has been replaced by other forms referred to as 'fiat money', which have gradually become the most common ones, such as metal coins and paper notes. Nowadays, other forms of money, such as electronic money, have become the most frequent form used to carry out transactions. In any case the most relevant points about money employed are its basic functions, which according to standard economic theory are:

- to serve as a medium of exchange, which is universally accepted in trade for goods and services;
- to act as a measure of value, making possible the determination of the prices and the calculation of costs, or profit and loss;
- to serve as a standard of deferred payments, i.e. a tool for the payment of debt or the unit in which loans are made and future transactions are fixed;
- to serve as a means of storing wealth not immediately required for use.

A related feature relevant for the present investigation is that money is the medium in which prices or values of all commodities as well as costs, profits and transactions can be determined or expressed. *Wealth* is usually understood as things that have economic utility (monetary value or value of exchange), or material goods or property; it also represents the abundance of objects of value (or riches) and the state of having accumulated these objects; for our purpose, it is important to bear in mind that wealth can be measured in terms of money. Also *income*, defined in Case and Fair (2008) as 'the sum of all the wages, salaries, profits, interests payments, rents and other forms of earnings received . . . in a given period of time', is a quantity which can be measured in terms of money (per unit time).[2]

2.2 Empirical analyses using data from earlier periods

It was first observed by Pareto (1897) that in an economy the higher end of the distribution of income $f(x)$ follows a power law,

$$P(x) \sim x^{-1-\alpha}, \tag{2.1}$$

[1] In *Encyclopædia Britannica*. Retrieved 18 June 2012 from Encyclopædia Britannica Online.
[2] See Chakraborti *et al.* (2011).

with α, now known as the Pareto exponent, estimated by him to be $\alpha \approx 3/2$. For the last hundred years the value of $\alpha \sim 3/2$ seems to have changed little in time and across the various capitalist economies (see Yakovenko and Barkley Rosser 2009, and references therein). The normalized Pareto distribution has the form

$$P(x) \sim \begin{cases} F(x) & \text{for } x < x_c, \\ \frac{\alpha x_c^\alpha}{x^{1+\alpha}} & \text{for } x \geq x_c, \end{cases} \tag{2.2}$$

where $F(x)$ is sometimes assumed to be $x^n \exp(-x/T)$ in the research communities; n and T are two constants. The above distribution has a scale x_c, which denotes a crossover from one kind of distribution to another, separating the low and middle wealth from the large wealth regime.

Gibrat (1931) clarified that Pareto's law is valid only for the high-income range, whereas for the middle-income range he suggested that the income distribution is described by a log-normal probability density

$$P(x) \sim \frac{1}{x\sqrt{2\pi\sigma^2}} \exp\left\{-\frac{\log^2(x/x_0)}{2\sigma^2}\right\}, \tag{2.3}$$

where $\log(x_0) = \langle\log(x)\rangle$ is the mean value of the logarithmic variable and $\sigma^2 = \langle[\log(x) - \log(x_0)]^2\rangle$ is the corresponding variance. The factor $\beta = 1/\sqrt{2\sigma^2}$, also known as the Gibrat index, measures the equality of the distribution. Empirically, β is known to lie between 2 and 3 (Souma 2002).[3]

Abul-Magd (2002) studied the wealth distribution in ancient Egypt. Excavations of the ancient Egyptian city Akhetaten, which was populated for only a brief period during the fourteenth century BC, have yielded a distribution of the house areas. Abul-Magd assumed that the house area is a measure of the wealth of its inhabitants, and made a comparative study of the wealth distributions in ancient and modern societies. According to his analysis of the wealth distribution in Akhetaten, the best-fit value of the Pareto exponent for the resulting distribution is $\alpha = 1.59 \pm 0.19$, which agrees very well with the values of the Pareto index obtained for modern societies.

Hegyi *et al.* (2007) also found a power law tail for the wealth distribution of aristocratic families in medieval Hungary. Hegyi *et al.* assumed that the number of serf families belonging to a noble is a measure of the corresponding wealth. He obtained a Pareto law for such a society with Pareto index $\alpha = 0.92$, which is smaller than the values reported for studies of the current period. The results obtained are plotted in Fig. 2.1.

[3] Historically speaking, Gibrat also analysed the firm size distribution and he proposed a 'law of proportionate effect'. This stated that a small change in a quantity is independent of the quantity itself. Thus, the distribution of a quantity $dz = dx/x$ should be Gaussian, and hence x as log-normal. A random variable is said to be log-normally distributed if its logarithm is normally distributed, also now known as *Gibrat's law*.

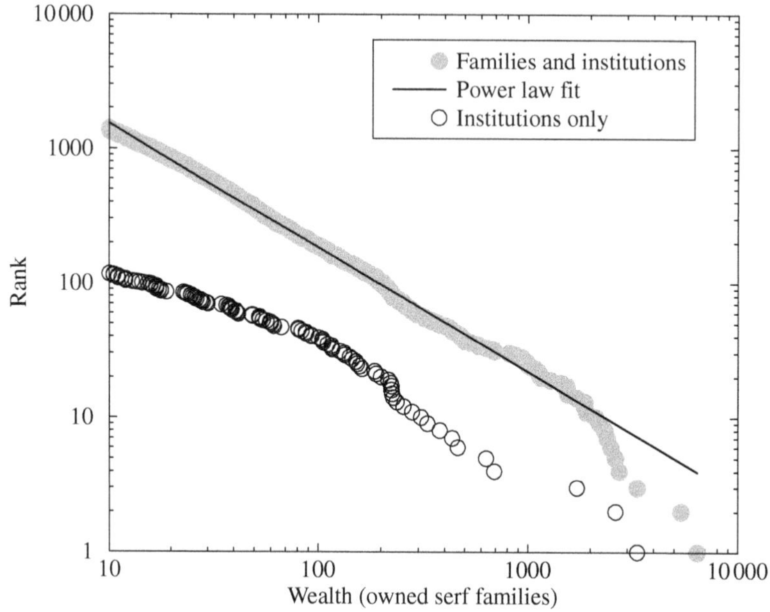

Figure 2.1 The rank of the top 8% aristocratic families and institutions as a function of their estimated total wealth on a double-logarithmic scale. Estimations made for the Hungarian noble society in the year 1550, in which the total wealth of a family is taken as the number of owned serf families. The power law fit suggests a Pareto index $\alpha = 0.92$. Reproduced from Hegyi *et al.* (2007).

Souma (2001) investigated the Japanese personal income distribution in the high-income range over the 112 years 1887–1998, and that in the middle-income range over the 44 years 1955–98. His data analysis revealed that the personal income followed the log-normal distribution (Gibrat) with a power law tail (Pareto). Since the same behaviour was observed in the analysis for the different years, it can be considered a statistical regularity. Figures 2.2 and 2.3 show the variations of the exponents that were obtained from the analyses of the log-normal distributions (for Gibrat index β) and the power law tails (for Pareto index α).

2.3 Empirical analyses using data from recent periods

According to Pareto (1897), 'the society is not homogeneous'. Hence, Pareto and many others were of the opinion that the distribution of income in a particular society would be an excellent indicator of non-homogeneity of any society. Thus, it would be interesting to study the empirical data and analyse the money, wealth and income distributions. Unfortunately, empirical data of money and wealth are rather scarce, and more data are available for the distribution of income from tax

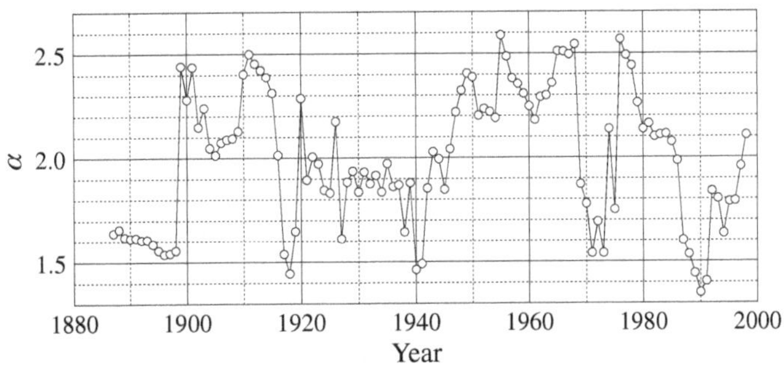

Figure 2.2 The variation of the Pareto index α over the 112 years 1887–1998. Reproduced from Souma (2001).

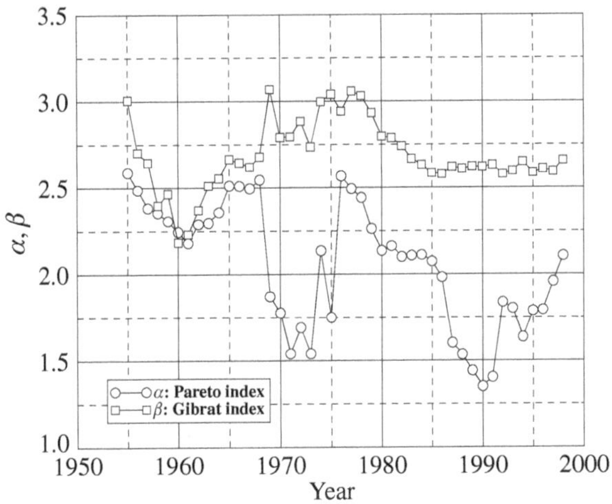

Figure 2.3 The variation of the Pareto index α and the Gibrat index β over the 44 years 1955–98. Reproduced from Souma (2001).

agencies and population surveys. Several analyses and studies have been done by economists and econometricians, and also recently by physicists. The approaches and terminologies have often been different, as have been the models and parametric fits of the data. In this section, we briefly present some representative empirical studies primarily done by physicists. We will often direct the readers to the original articles and other reviews which are more exhaustive.

For money, in principle, it should have been easy to study its distribution. Since most people keep their money in banks, one could approximate the distribution of money by the distribution of balances on all types of bank accounts, where the data for a very large bank would be representative of the distribution in the entire economy. However, such data of clients are not made available to academic researchers, either by the banks or by organizations such as the Federal Deposit Insurance Corporation of the USA, which insures bank deposits of customers up to a certain maximal balance, as already remarked by Yakovenko and Barkley Rosser (2009).

For wealth, again it is quite difficult – in most countries, it is not something that is officially reported by individuals to the authorities. Yakovenko and Barkley Rosser (2009) observed that in a few countries, when a person dies, all assets are required to be reported for the purpose of 'inheritance tax'. Hence, ironically there exist some good statistics of wealth distribution among 'dead' people, rather than the wealth distribution among the 'living'. For example, using certain statistical methods and adjustment procedures based on several factors such as age, gender and other characteristics of the dead people, the tax agency of the UK – the Inland Revenue – was able to reconstruct the wealth distribution of the whole population of the UK.[4] These data were studied by Drăgulescu and Yakovenko (2001b), reported later in the section.

Income distribution has been much easier to study. The empirical studies of income distribution have a long history in the economics and econometrics literature (Kakwani 1980; Champernowne and Cowell 1998; Atkinson and Bourguignon 2000; Piketty and Saez 2003; Atkinson and Piketty 2007). Income distributions have been studied extensively in econophysics papers, for many different countries: Australia (Di Matteo *et al.* 2004; Banerjee *et al.* 2006; Clementi *et al.* 2006, 2008), Germany (Clementi and Gallegati 2005a; Clementi *et al.* 2007), India (Sinha 2006), Italy (Clementi and Gallegati 2005b; Clementi *et al.* 2006, 2007), Japan (Souma 2001, 2002; Aoyama *et al.* 2003; Fujiwara *et al.* 2003; Ferrero 2004, 2005; Souma and Nirei 2005; Nirei and Souma 2007), New Zealand (Ferrero 2004, 2005), the UK (Ferrero 2004, 2005; Clementi and Gallegati 2005a; Richmond *et al.* 2006; Clementi *et al.* 2007) and the USA (Drăgulescu and Yakovenko 2001a,b; Drăgulescu and Yakovenko 2003; Rawlings *et al.* 2004; Clementi and Gallegati 2005a; Clementi *et al.* 2008).

The wealth and income distributions are qualitatively very similar, and there appears to be a statistical regularity: the upper tail follows the power law (Pareto), and comprises a small fraction of population; the lower part of the distribution follows one of the exponential (Gibbs) or gamma or log-normal (Gibrat) distributions. The studies (Souma 2001, 2002) were made to investigate the lower part of the distribution: the log-normal or *Gibrat law*, which we mentioned earlier. A few other studies and statistical surveys of the population, such as the Survey of Consumer Finance (Diaz-Giménez and Rios-Rull 1997) and the Panel Study of Income

[4] HM Revenue & Customs, 2003. Distribution of Personal Wealth. http://www.hmrc.gov.uk/stats/personal_wealth/menu.htm.

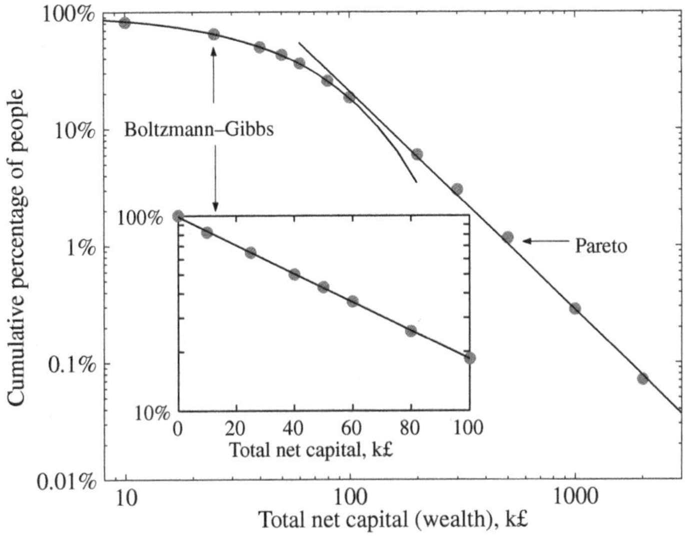

Figure 2.4 The cumulative probability distribution of net wealth in the UK in 1996 shown on log-log (main panel) and log-linear (inset) scales. Points represent the data from the Inland Revenue (tax agency), and solid lines are the fitted lines to the exponential (Boltzmann–Gibbs) and power (Pareto) laws. Reproduced from Drăgulescu and Yakovenko (2001b).

Dynamics (PSID), have also shed some light on the lower part of the distribution. There have been several studies of the *Pareto* power law by physicists (Aoyama *et al.* 2000; Drăgulescu and Yakovenko 2001a,b, 2003; Levy 2003; Levy and Levy 2003; Sinha 2006; Klass *et al.* 2007) for the upper-tail data.

Incidentally, the consumer expenditure distribution also indicates a mixture of the log-normal and Pareto distributions, with the Pareto tail comprising approximately 10–20% of the population (Mizuno 2008; Mizuno *et al.* 2008; Ghosh *et al.* 2009). In this book, we restrict ourselves to the detailed discussions of wealth and income distributions only.

Drăgulescu and Yakovenko (2001b) studied the UK wealth data for 1996. Figure 2.4 shows the cumulative probability $C(w) = \int_w^\infty P(w')\,dw'$ plotted as a function of the personal net wealth w (composed of assets, such as cash, stocks, property, household goods, and liabilities, such as mortgages and other debts). The main panel shows a plot of $C(w)$ on the double-logarithmic scale, where the straight line indicates a power law dependence:[5] $C(w) \propto 1/w^\alpha$ with the exponent $\alpha = 1.9$ for wealth greater than approximately 100 k£. The inset shows the same data on

[5] Whenever the statistical data are usually reported at non-uniform intervals, it is often more practical to plot the cumulative probability distribution rather than its derivative, the probability density. Interestingly, when the probability density $P(x)$ is an exponential or a power law function, the cumulative probability distribution $C(x)$ is also an exponential or a power law function.

the log-linear scale, where a straight line indicates an exponential behaviour. It was observed that, below 100 k£, the data fit well the exponential distribution $C(w) \propto \exp(-w/T_w)$ with the effective 'wealth temperature' $T_w = 60$ k£ (corresponding to a median wealth of 41 k£). Therefore, the distribution of wealth may be characterized by the two parts: the Pareto power law in the upper tail of the distribution, and the exponential Boltzmann–Gibbs law in the lower part of the distribution for the great majority (approximately 90%) of the population (Drăgulescu and Yakovenko 2001b). Yakovenko and Barkley Rosser (2009) suggested that wealth distribution in the lower part is dominated by distribution of money, because the corresponding people do not have other significant assets (Levy and Levy 2003), resulting in the Boltzmann–Gibbs law; whereas the upper tail of wealth distribution is dominated by investment assets (Levy and Levy 2003), resulting in the Pareto law. Models and mechanisms resulting in the above distributions will be discussed in the following chapters in more detail.

Klass *et al.* (2007) used in their study the list that *Forbes* magazine publishes once a year: the list of the 400 richest people in the USA.[6] The list includes people from diverse backgrounds and asserts the net worth of each individual. What is interesting is the fact that even though the people included in the *Forbes* 400 list have gathered their fortunes in myriad ways, there exists a striking statistical regularity in the distribution of their wealth! Analysing the *Forbes* 400 lists during the period 1988–2003, Klass *et al.* found the statistical regularity that, at the top end of the wealth distribution, the wealth is distributed according to a Pareto distribution. The wealths w_r of the 400 richest Americans in 2003, ordered by their ranks r, are shown on a double-logarithmic scale in Fig. 2.5. The data on this plot, known as the *Zipf plot* (Newman 2005), can be fitted by a straight line, which indicates that the wealths exhibit a power law behaviour of the form

$$w_r \sim r^{-\gamma},$$

where the exponent $\gamma = 0.78 \pm 0.05$ is called the *Zipf exponent*. The above wealth–rank relation implies a power law distribution of the wealth

$$P(w) \sim w^{-(\alpha+1)},$$

which is nothing but the Pareto distribution, and α is the Pareto exponent. However, because of the relatively small number of data points for a particular year, the resulting distribution $P(w)$ obtained after binning the data turns out to be somewhat noisy, and so, for a single year, the Zipf plot provides more reliable results. For the general connection between the Pareto and the Zipf exponents: $\alpha = 1/\gamma$, please

[6] www.forbes.com/lists.

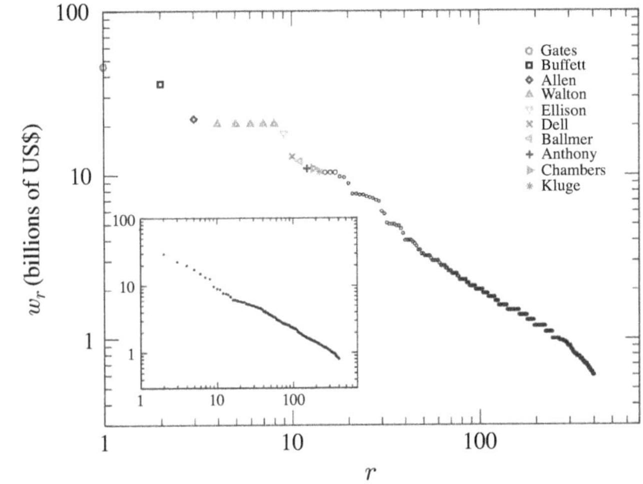

Figure 2.5 Zipf plot of the wealths w_r of the people in the *Forbes* 400 list in the year 2003 vs their ranks r. The power law fit, with the Zipf exponent $\gamma = 0.78 \pm 0.05$, which was obtained in the range $10 \leq r \leq 300$. Adapted from Klass *et al.* (2007).

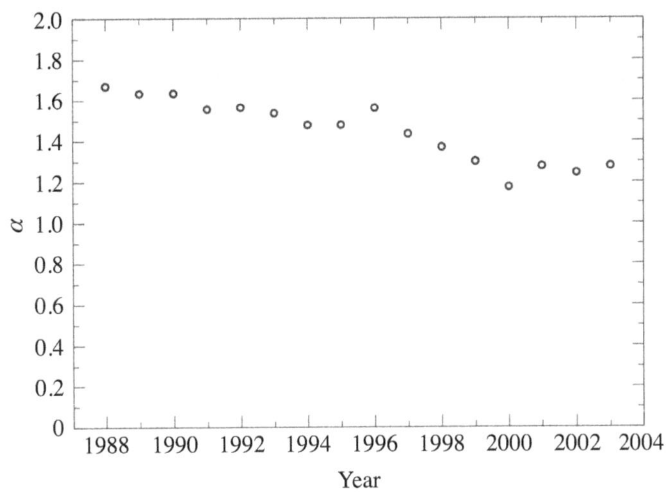

Figure 2.6 The Pareto exponent $\alpha(t)$, obtained from the Zipf analysis, using the relation $\alpha = 1/\gamma$, vs the year t, during 1988–2003. The power law fitting of the Zipf data was obtained using the range $10 \leq r \leq 300$. Adapted from Klass *et al.* (2007).

see Newman (2005). Therefore, the power law distribution in Fig. 2.5 actually corresponds to $\alpha = 1.28$. To examine the temporal variations of α, repeated Zipf analysis was performed for each year during 1988–2003. It was found that α varies widely in the range between 1.1 and 1.7, as shown in Fig. 2.6.

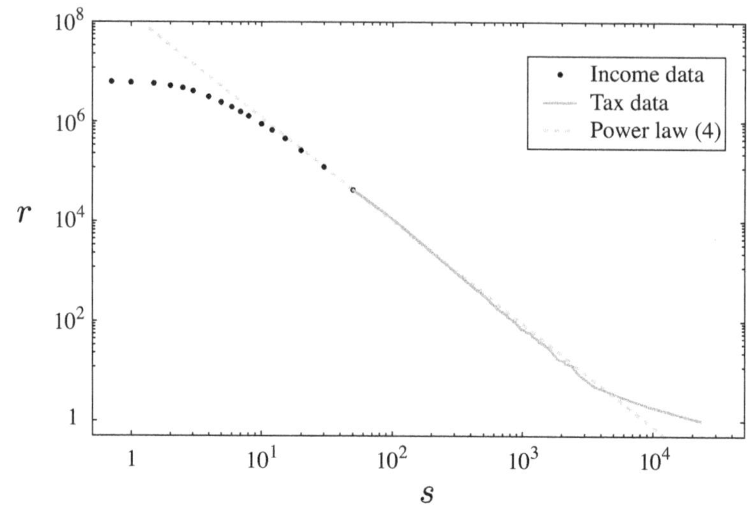

Figure 2.7 The rank-size Zipf plot of the income. The income data are shown by dots, whereas the data from the income tax are connected by the solid line segments. The broken line shows the fitted line. Reproduced from Aoyama *et al.* (2000).

Aoyama *et al.* (2000) had earlier analysed the distributions of the income and income tax of individuals in Japan. The income data were for the fiscal year 1997 and 1998, whereas the income tax data were for only 1998. For the fiscal year 1998, the income data contained all 6 224 254 workers who filed tax returns, but they were a coarsely tabulated data; the income tax data listed the income tax of individuals who paid tax of 10 million yen or more in the same year. They studied the individual distribution of each data set and then combined them carefully to obtain an overall picture of income distribution in the high-income range. The resulting rank-size Zipf plot, which is shown in Fig. 2.7, obeys a power law with a Pareto exponent very close to -2.

The time-evolution of the two-part structure of the empirical income distribution was studied by Clementi and Gallegati (2005a). Figure 2.8 shows that the same structure holds for the entire time spans studied, and for all the three countries: the USA (1980–2001), Germany (1990–2002) and the UK (1991–2001). They had used the income data from the US PSID, the British Household Panel Survey and the German Socio-Economic Panel, as released in a cross-nationally comparable format in the Cross-National Equivalent File (CNEF).[7]

For distributional analysis, fitting a parametric model using such functions to the income data is a valuable and informative tool, since one can not only characterize

[7] See the CNEF web site for details: http://www.human.cornell.edu/pam/research/centers-programs/german-panel/cnef.cfm.

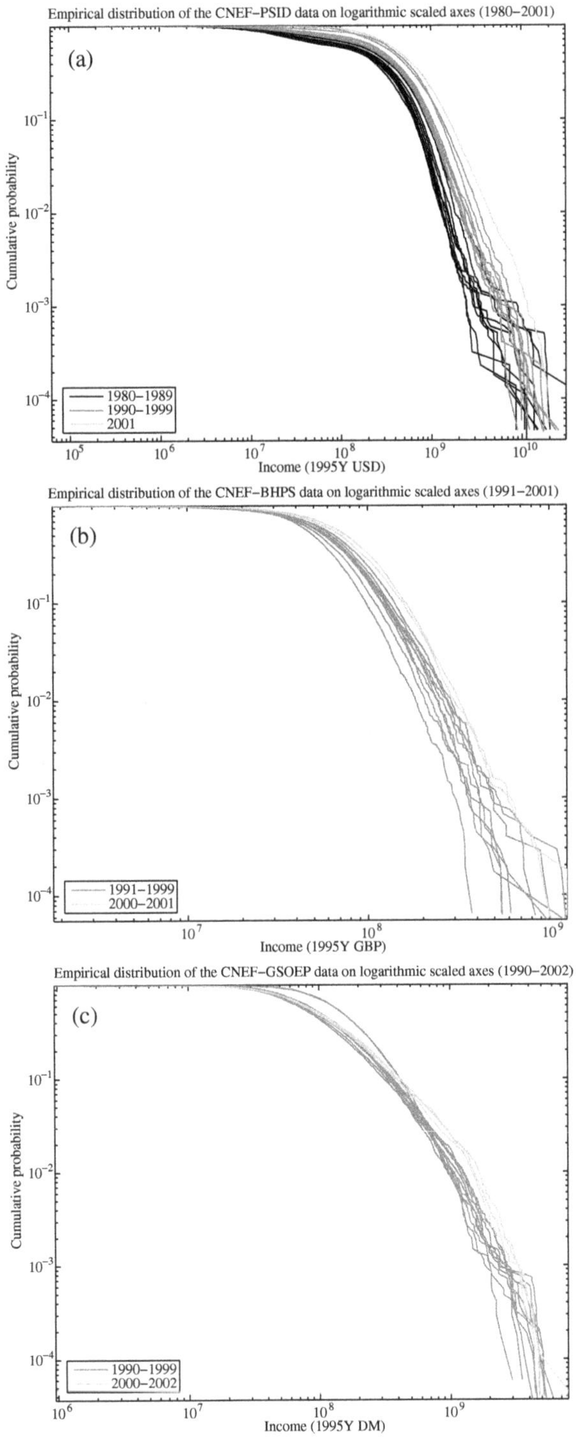

Figure 2.8 (a–c) Time developments of income distributions for the USA (1980–2001), Germany (1990–2002) and the UK (1991–2001). Reproduced from Clementi and Gallegati (2005a). BHPS, British Household Panel Survey; CNEF, Cross-National Equivalent File; GSOEP, German Socio-Economic Panel; PSID, Panel Study of Income Dynamics.

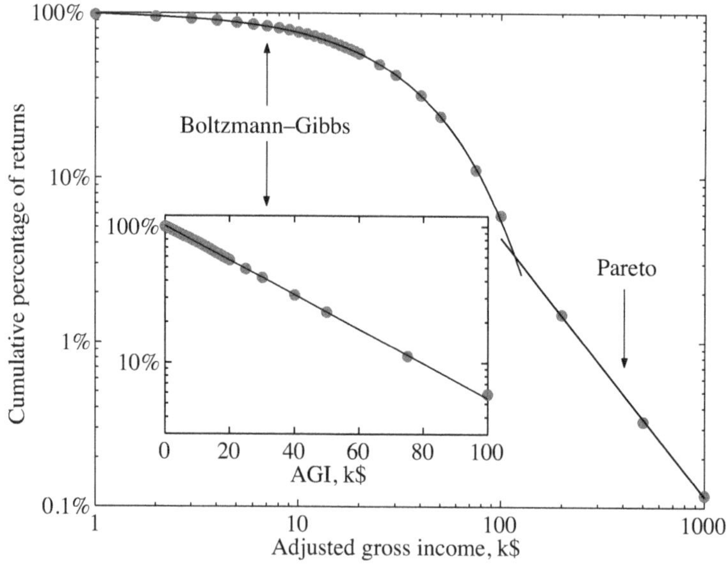

Figure 2.9 Plot of the cumulative probability distribution of tax returns for the USA in 1997 shown on log-log (main panel) and log-linear (inset) scales. Points represent the Internal Revenue Service (tax agency) data, and solid lines are fitted lines to the exponential and power law functions. Reproduced from Drăgulescu and Yakovenko (2003).

the information contained in the numerous observations, but also draw practical information directly from the estimated parameters. For example, one could be interested in measuring income inequality and comparing different distributions – these concepts may be directly derived from parameters of a fitted distribution. Some measures of inequality and comparison of distributions of income from different countries are discussed in the last section of this chapter.

Drăgulescu and Yakovenko (2001a,b, 2003) also made an interesting observation that the income distribution can be fitted in two parts: an exponential function in the lower part – $P(r) = c\exp(-r/T_r)$ – characterized by the 'income temperature' T_r, and a power law function in the upper part, as shown in Fig. 2.9. As in the earlier case of wealth studies, the straight line on the log-linear scale in the inset demonstrates the exponential *Boltzmann–Gibbs* law, and the straight line on the double-logarithmic scale in the main panel shows the *Pareto* power law. It was also suggested that the fact that income distribution consists of two distinct parts actually reveals the 'two-class structure' in the society (Silva and Yakovenko 2005; Yakovenko and Silva 2005). The similar coexistence of exponential and power law distributions was known much earlier in plasma physics and astrophysics, in which they were called the 'thermal' and 'superthermal' parts

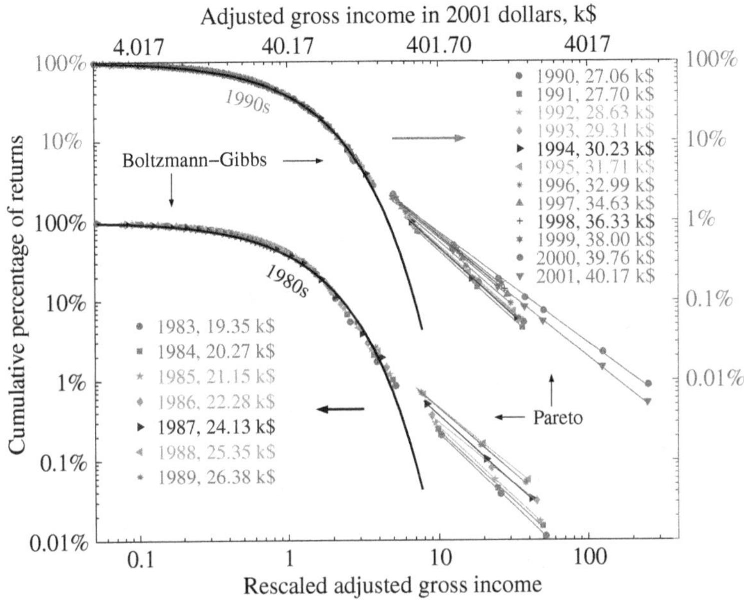

Figure 2.10 Plot of the cumulative probability distribution of tax returns plotted on log-log scale versus r/T_r (the annual income r normalized by the average income T_r in the exponential part of the distribution). The Internal Revenue Service data points are for 1983–2001, and the columns of numbers give the values of T_r for the corresponding years. Reproduced from Silva and Yakovenko (2005).

(Hasegawa *et al.* 1985; Desai *et al.* 2003; Collier 2004). The boundary between the lower and upper classes can be defined as the crossover point of the exponential (thermal) and power law (superthermal) fits, as shown in Fig. 2.9. As Drăgulescu and Yakovenko (2003) pointed out, for 1997, the annual income separating the two classes was approximately 120 k$, such that approximately 3% of the population belonged to the upper class and the remaining 97% belonged to the lower class.

Later, Silva and Yakovenko (2005) studied the time evolution of income distribution in the USA during 1983–2001 using the data from the Internal Revenue Service (IRS), the government tax agency for the USA. A remarkable regularity was observed: the structure of income distribution was qualitatively similar for all years, as evident in Fig. 2.10. Note that, since the average income in nominal dollars approximately doubled during this time interval, the normalized income r/T_r, where the 'income temperature' T_r was obtained by fitting the exponential part of the distribution for each year, is plotted in the horizontal axis in Fig. 2.10. The values of T_r are shown in Fig. 2.10. The plots for the 1980s and 1990s are adjusted on the vertical axes for clarity. What is intriguing is the different behaviour in the time evolution of the lower and upper parts of the income distribution. They observed that the data points in the lower-income part of the distribution collapsed on the same exponential curve for all years, which suggests the fact that the shape of the income distribution for the lower class is very robust and does not evolve with time, despite the gradual increase in the average income in nominal dollars. This observation leads to the analogy of a statistical 'thermal' equilibrium in the lower-class income distribution. Figure 2.10 also shows that the income distribution of the upper class does not rescale at all, and instead evolves significantly in time. Silva and Yakovenko (2005) found that the exponent α of the power law $C(r) \propto 1/r^{\alpha}$ decreased from 1.8 in 1983 to 1.4 in 2000, which meant that the upper tail actually became 'fatter'.

Silva and Yakovenko (2005) also studied the behaviour of another parameter, f, the ratio of the total income of the upper class and the total income in the system, which they found increased drastically from 4% in 1983 to 20% in 2000. However, in 2001, α increased and f decreased, suggesting that the upper tail was reduced after the stock market crash which occurred then. These results show that the upper tail is not stationary, and might be a good indicator of the overall economy: it tends to swell during the stock market bubble and shrink during the burst.

We now present another case study, in which Ferrero (2004) found evidence that the income distribution for different countries followed a gamma function of the form

$$P(x) \sim x^{n-1} \exp(x/T),$$

which can be naturally associated to the Boltzmann distribution of energy in polyatomic molecules instead of the simple exponential of an ideal monatomic gas. The data were obtained from the statistical information provided by the revenue services of Japan, the UK and New Zealand.[8] Since the available data were given in different-sized bins, they were reprocessed to obtain the 'normalized count' – defined as the count in the class divided by the number of observations times the class width.[9] Figure 2.11 shows the empirical income distribution for Japan, New Zealand and the UK, where the money scale is given in thousands of New Zealand dollars, and in order to show the distributions from the three countries in the same graph, the data corresponding to the UK were divided by 2 and those from Japan by 2000. The fitted values of the parameters n and T are given in Table 2.1.

Note that the available data were provided by revenue services or statistical offices of different countries as individual income distributions, and therefore a necessary assumption in this study was that the money actually possessed by the

[8] Aoyama *et al.* (2000); Inland Revenue: income distribution (http://www.hmrc.gov.uk/stats/income_distribution/menu-by-year.htm); New Zealand Income Survey (http://www.stats.govt.nz/browse_for_stats/income-and-work/income.aspx).

[9] Engineering Statistics Handbook http://www.itl.nist.gov/div898/handbook.

Table 2.1 *Parametric values of* n *and* T *obtained from the fits of the income data for New Zealand (1998), the UK (1998–99) and Japan (1996)*

	n	T
New Zealand	1.67	12.34 kNZ$
UK	4.28	5.11 k£
Japan	3.66	2.63 MJPY

Values taken from Ferrero (2004).

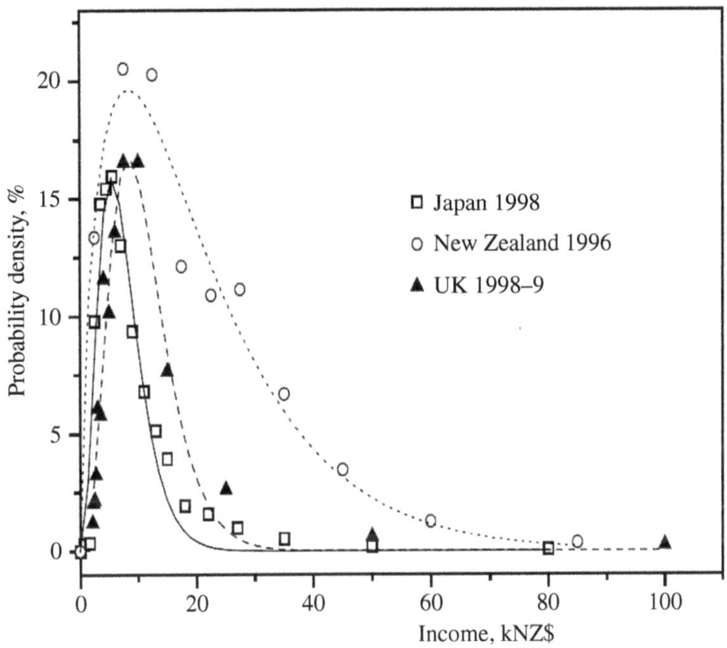

Figure 2.11 Income distribution for Japan, New Zealand and the UK. The income values for Japan and the UK were rescaled to show in the same plot. The lines represent the best fit to the gamma function, as indicated in the text. Adapted from Ferrero (2004).

agents is proportional to the income (which cannot be completely true). There were a few more theoretical assumptions that made possible the analogy between energy exchange of gas and money exchange of agents, but we do not discuss them now.

Interestingly, some papers have also used interpolating functions with different asymptotic behaviour for low and high incomes, such as the Tsallis function (Ferrero 2005) or the Kaniadakis function (Clementi *et al.* 2007, 2008). We discuss

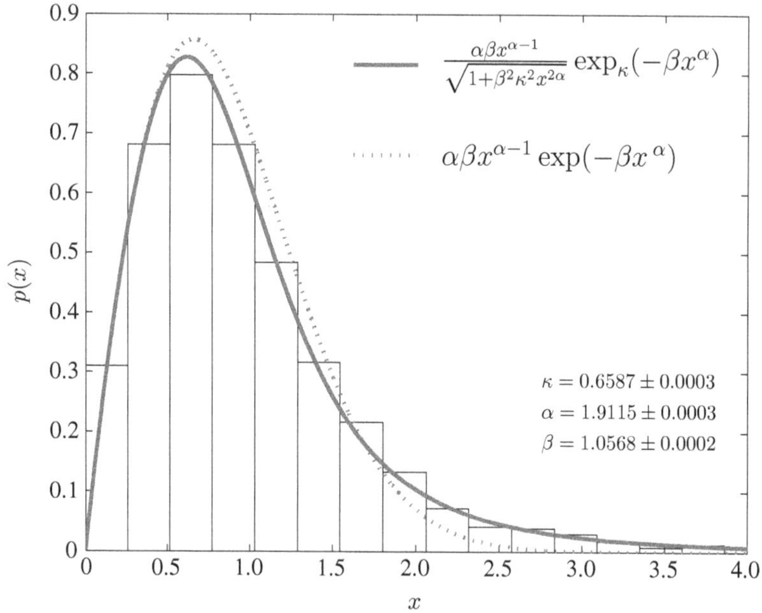

Figure 2.12 The mean-rescaled US personal income distribution in 2003. Probability density histogram with superimposed fits of the κ-generalized (solid line) and Weibull (dotted line) densities. The income axis limits have been adjusted according to the range of data to shed light on the intermediate region between the bulk and the upper end of the distribution. Reproduced from Clementi *et al.* (2008).

another example of a parametric approach to income analysis and comparison with real data, following Clementi *et al.* (2007, 2008). In these papers, they proposed a new fitting function having its roots in the framework of the κ-generalized statistical mechanics:

$$p(x) = \frac{\alpha\beta x^{\alpha-1}\exp_\kappa(-\beta x^\alpha)}{\sqrt{1+\kappa^2\beta^2x^{2\alpha}}}. \tag{2.4}$$

The model distribution has a bulk which is very close to the stretched exponential one – which is recovered when the deformation parameter κ tends to zero – while for high values of income, the upper tail of the distribution approaches a Pareto distribution. Hence, it is able to fit the data over the entire range.

The performance of the distribution was checked against real data. The κ-generalized distribution was fitted to data on personal income derived from the 2003 wave of the US PSID as released in the CNEF – a commercially available database compiled by researchers at Cornell University.[10] The 2003 PSID-CNEF

[10] See the CNEF web site for details: http://www.human.cornell.edu/pam/gsoep/equivfil.cfm.

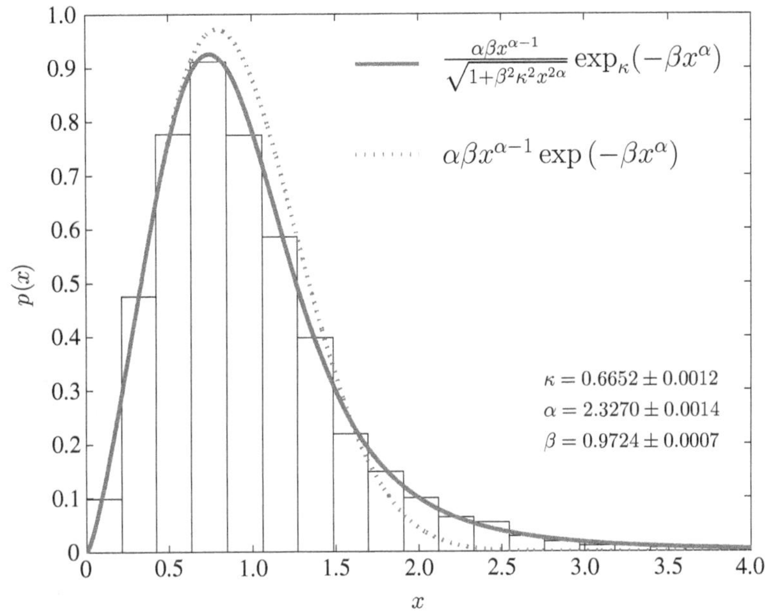

Figure 2.13 The Australian personal income distribution in 2002–3 measured in current year AUD. Probability density histogram with superimposed fits of the κ-generalized (solid line) and Weibull (dotted line) densities. The income axis limits have been adjusted according to the range of data to shed light on the intermediate region between the bulk and the upper end of the distribution. Reproduced from Clementi *et al.* (2008).

data had a sampling of 7822 households, and all calculations were based on the household post-government income – the income recorded after taxes and government transfers – expressed in nominal local currency units, and normalized to its empirical average. The best-fitting parameter values were determined using ‘constrained maximum likelihood’ estimation, resulting in the following estimates: $\alpha = 1.9115 \pm 0.0003$, $\beta = 1.0568 \pm 0.0002$ and $\kappa = 0.6587 \pm 0.0003$. The very small value of the errors indicates that the parameters were precisely estimated, and the comparison between the observed and fitted probabilities in Fig. 2.12 shows that the data are remarkably well fitted. The κ-generalized distribution was also fitted to data on personal income distribution for Australia (Clementi *et al.* 2008). The data were derived from panel surveys conducted in 2002–3. The unit of assessment was the household, and income was expressed in nominal local currency units for the 10 211 households in the 2002–3 Australian survey. Similar analysis was conducted to yield Fig. 2.13.

It must be mentioned that, for example, for the US data shown in Figs. 2.9 and 2.10, the transition between the lower and upper classes is not always very smooth.

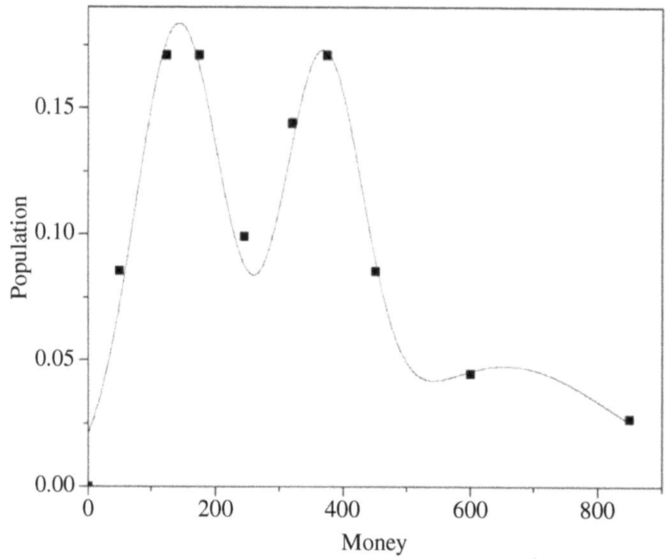

Figure 2.14 The income distribution of Argentina in May 2002. Reproduced from Ferrero (2005).

In such cases, the fitting using interpolating functions, such as above, would not be useful.

One must also mention that many studies have found special features:

(1) Ferrero (2005) studied the income distribution in Argentina during the economic crisis, which showed a time-dependent bimodal shape with two peaks (Fig. 2.14).
(2) Coelho *et al.* (2008) examined wealth distributions and found the existence of two distinct power law regimes: the Pareto exponents of the super-rich (identified, for example, in rich lists such as that provided by *Forbes*) are smaller than the Pareto exponents obtained for top earners in income data sets. Their results were based on the studies for income and wealth distributions around the world (see Richmond *et al.* 2006, Table 5.2). Figure 2.15 displays the distribution of the Pareto exponents, and it is evident that, while the average Pareto exponent is approximately 2.0 for the top earners in income tax/inheritance statistics, it is just below 1.0 for the super-rich. Their explanation was that the studies of wealth that are based on tax/income generally do not include the wealth of very rich people, and there were several such instances. An instance of two power law regimes was the study of Souma (2001), who found a Pareto exponent of 2.06 in the high end. However, there is an indication of a second power law for the top richest (higher than 3000 million yen) which has an exponent below 1.0 based on his figure, as shown in Fig. 2.16.

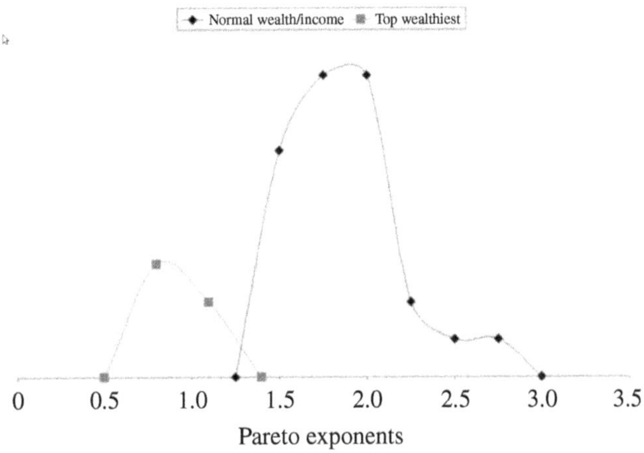

Figure 2.15 Distribution of the Pareto exponents found by different authors in the last decade. The black curve is from data sets taken from tax/income databases. The grey curve is from super-rich lists, such as *Forbes*. The Pareto exponent for the top richest is around 1, while for the 'normal' rich people it is around 2 (data taken Richmond *et al.* 2006, Table 5.2). Reproduced from Coelho *et al.* (2008).

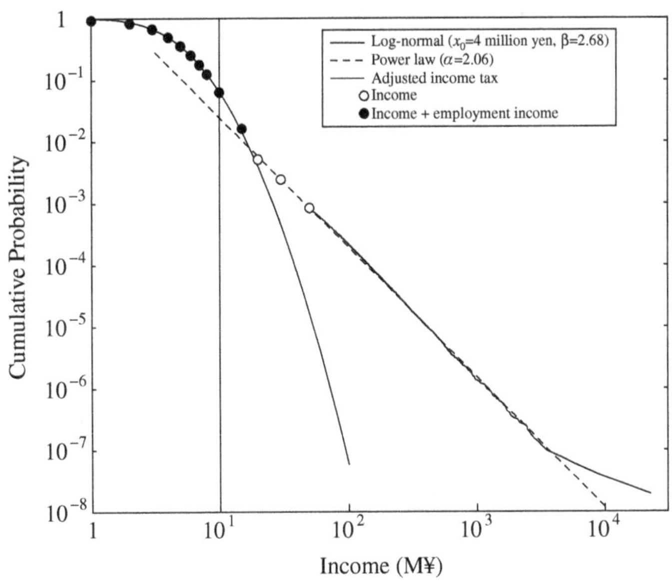

Figure 2.16 The cumulative probability of Japanese personal income in 1998. Reproduced from Souma (2001).

(3) Coelho *et al.* (2008) gave another instance of two power laws in their analysis of UK data. Figure 2.17 shows data for the cumulative distribution of incomes in the UK for the year 1995. In one region the Pareto exponent is ~3.3. Then there is a second power law with Pareto exponent ~1.26. They think that this

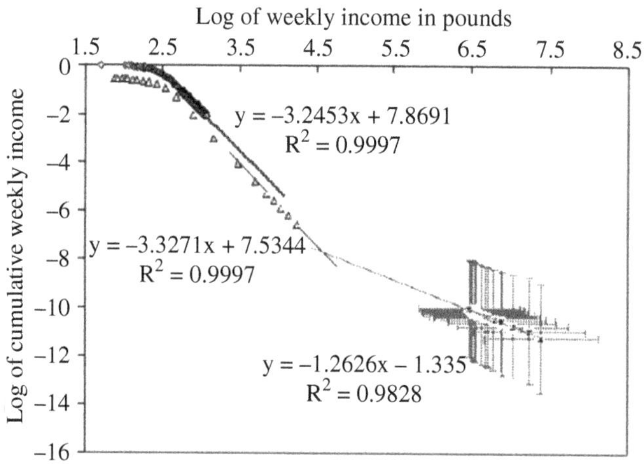

Figure 2.17 Distribution of the cumulative weekly income in the UK for 1995. The left-hand curves represent income for 1995 from two different sources and a similar Pareto exponent is achieved for the high end of these curves, ~3.2–3.3. The right-hand curve represents an estimation of the income, in 1995, for the top richest in the UK. In this case the Pareto exponent is lower and around 1.3. Reproduced from Coelho *et al.* (2008).

corroborates the belief of many people that the super-wealthy pay less tax as a proportion of their income than the majority of earners in society!

Finally, one must remark that, in reality, the income distributions are more often reported by statistical agencies for households, and so it is always very difficult to differentiate between one-earner and two-earner income distributions. Hence, after discussing the distribution of individual income, it is a matter of interest to study the related distribution $P_2(r)$ of family income $r = r_1 + r_2$, where r_1 and r_2 are the incomes of spouses. Note that, if the individual incomes are distributed exponentially $P(r) \propto \exp(-r/T_r)$, then from simple mathematics it follows that

$$P_2(r) = \int_0^r dr' P(r')P(r - r') = c\, r \exp(-r/T_r), \tag{2.5}$$

where c is a normalization constant, assuming that that there are no correlations in the incomes of the spouses. Drăgulescu and Yakovenko (2001a) showed that Eq. (2.5) is in good agreement with the family income distribution data from the US Census Bureau, as shown in Fig. 2.18. The assumption that the incomes of spouses are uncorrelated, made in Eq. (2.5), is indeed supported by the scatter plot of incomes of spouses (Fig. 2.19), in which each family is represented by two points (r_1, r_2) and (r_2, r_1) for symmetry. Drăgulescu and Yakovenko (2001a) observed that the density of points is approximately constant along the lines of

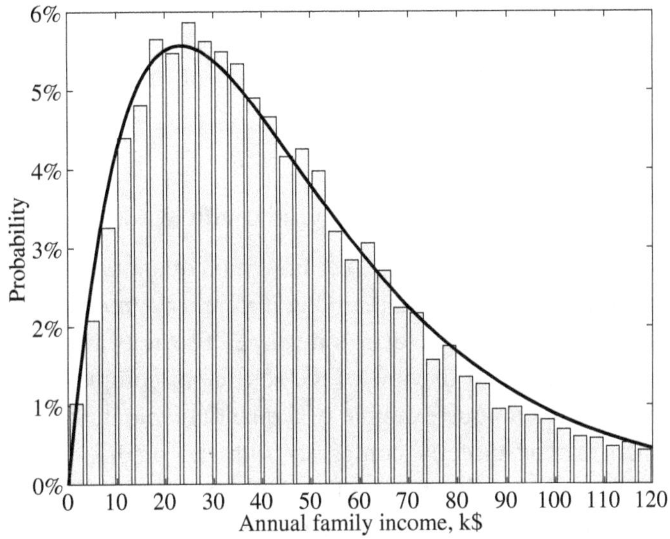

Figure 2.18 Histogram of the family income for families with two adults (US Census Bureau data for 1996). The solid line is the fit of the theoretical result given by Eq. (2.5). Reproduced from Yakovenko and Barkley Rosser (2009).

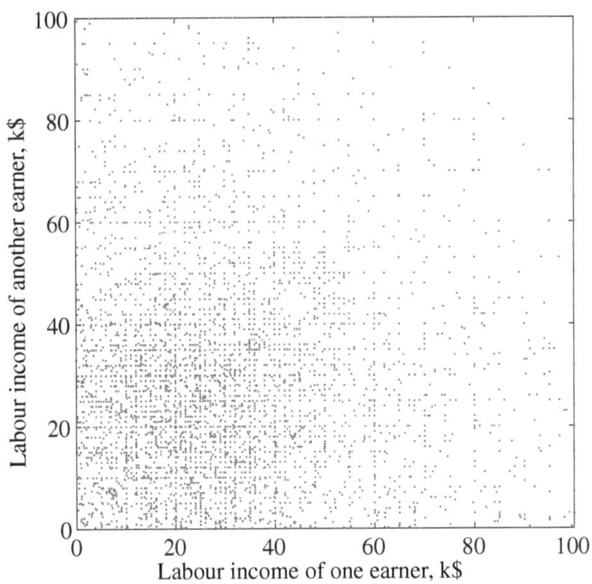

Figure 2.19 Scatter diagram of spouses' incomes (r_1, r_2) and (r_2, r_1) (symmetric), based on data from the Panel Study of Income Dynamics, 1999, explained in the text. Reproduced from Drăgulescu and Yakovenko (2003).

constant family income $r_1 + r_2 = \text{const.}$, which shows that the incomes of spouses are uncorrelated, approximately. The fact that there is no significant clustering of points along the diagonal $r_1 = r_2$ corroborates that there is no strong positive correlation of incomes of spouses.

2.4 Measures of income inequality: Gini coefficient and Lorenz curve

The distributions of wealth or income, i.e. how such quantities are shared among the population of a given country and among different countries, is a topic which has been studied by economists for a long time. The relevance of the topic to us is twofold. From the point of view of the science of complex systems, wealth distributions represent a unique example of a quantitative outcome of a collective behaviour which can be directly compared with the predictions of theoretical models and numerical experiments, which shall be described in the chapters to follow. Also, there is a basic interest in wealth distributions from the social point of view, in particular in their degree of (in)equality. To this aim, the Gini coefficient (or the Gini index, if expressed as a percentage), developed by the Italian statistician Corrado Gini, represents a concept commonly employed to measure inequality of wealth distributions or, more in general, how uneven a given distribution is (Gini 1921). For a cumulative distribution function $F(y)$ that is piecewise differentiable, has a finite mean μ and is zero for $y < 0$, the Gini coefficient is defined as

$$\begin{aligned} G &= 1 - \frac{1}{\mu}\int_0^\infty dy\,(1 - F(y))^2 \\ &= \frac{1}{\mu}\int_0^\infty dy\, F(y)(1 - F(y)). \end{aligned} \tag{2.6}$$

It can also be interpreted statistically as half the relative mean difference. Thus, the Gini coefficient is a number between 0 and 1, where 0 corresponds with perfect equality (where everyone has the same income) and 1 corresponds with perfect inequality (where one person has all the income, and everyone else has zero income).

The Gini coefficient can also be calculated from the Lorenz curve, which is a standard way of representing income distribution in the economic literature (Kakwani 1980). For a discussion of income inequality, one of the standard practices is to adopt the concept of concentration of incomes as defined by Lorenz 1905. Mathematically, the Lorenz curve is defined in terms of two coordinates $x(m)$ and $y(m)$ depending on a parameter m:

$$x(m) = \int_0^m P(m')\,dm', \quad y(m) = \frac{\int_0^m m'P(m')\,dm'}{\int_0^\infty m'P(m')\,dm'}, \tag{2.7}$$

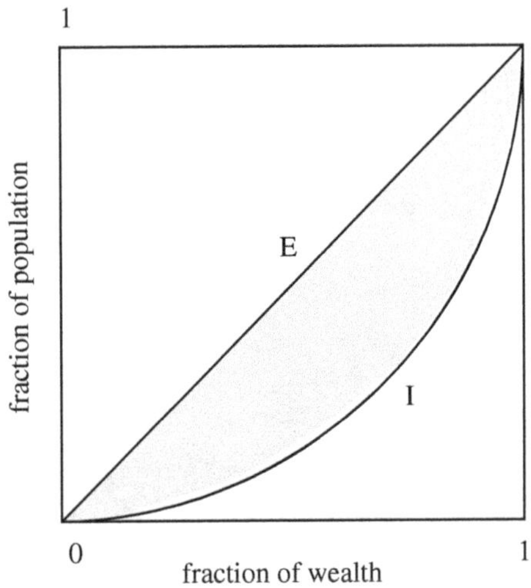

Figure 2.20 The Gini coefficient G gives a measure of inequality in any income distribution and is defined as the proportional area between I, giving the cumulative fraction of the people with the fraction of wealth, and the perfect equality curve E, where the fraction of wealth possessed by any fraction of population would be strictly linear: $G = 1 - \frac{A_I}{A_E}$, where A_I and A_E are the areas under curves I and E, respectively. $G = 0$ corresponds to perfect equality while $G = 1$ corresponds to perfect inequality.

where the horizontal coordinate $x(m)$ is the fraction of the population with income below m, and the vertical coordinate $y(m)$ is the fraction of the income which this population accounts for. Of course, as the parameter m changes from 0 to ∞, the coordinates x and y change from 0 to 1, and thus parametrically define a curve in the (x, y) plane (Yakovenko and Barkley Rosser 2009). Thus, if everybody had the same income, the Lorenz curve would be the diagonal line, because the fraction of income would be proportional to the fraction of the population, as illustrated in Fig. 2.20. The deviation of the Lorenz curve from the straight diagonal line in Fig. 2.20 is thus a certain measure of income inequality. As mentioned above, the standard measure of income inequality (Kakwani 1980) is the Gini coefficient $0 \leq G \leq 1$, and it may also be defined as the ratio of the area between the Lorenz curve and the diagonal line, and the area of the triangle beneath the diagonal line, illustrated in Fig. 2.20.

Empirical studies by Drăgulescu and Yakovenko (2001a) reveal that, even though the relative income inequality within the lower class remains stable, the overall income inequality in the USA has increased significantly as an outcome of the tremendous growth of the upper class income. Figure 2.21 shows the data points for the Lorenz curves in 1983 and 2000, as computed by the IRS (Strudler *et al.* 2003).

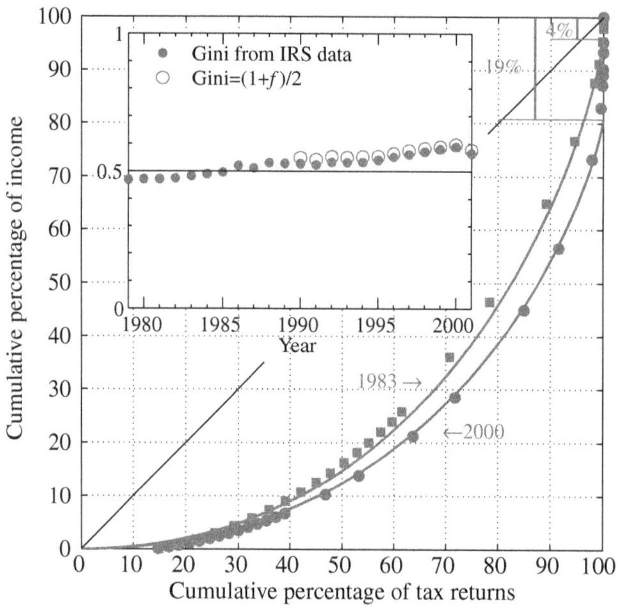

Figure 2.21 Lorenz plots for income distribution in 1983 and Internal Revenue Service 2000. The data points are from the (IRS) (Strudler *et al.* 2003), and the theoretical curves represent Eq. (2.9) with the parameter f deduced from Fig. 2.10. *Inset:* The closed circles are the IRS data (Strudler *et al.* 2003) for the Gini coefficient G, and the open circles show the theoretical formula $G = (1 + f)/2$. Reproduced from Silva and Yakovenko (2005).

For a purely exponential distribution, the Lorenz curve is

$$y = x + (1 - x)\ln(1 - x), \tag{2.8}$$

as shown by the upper curve in Fig. 2.21, which reasonably agrees with the 1983 data. Elsewhere (Drăgulescu and Yakovenko 2003; Silva and Yakovenko 2005; Yakovenko and Silva 2005), the authors suggested that one needs to take into account the fraction f of income in the upper tail, and suitably modify the Lorenz formula as

$$y = (1 - f)[x + (1 - x)\ln(1 - x)] + f\,\Theta(x - 1), \tag{2.9}$$

where the last term in Eq. (2.9) represents the vertical jump of the Lorenz curve at $x = 1$. At this point, a small percentage of population in the upper class actually accounts for a substantial fraction f of the total income. The lower curve in Fig. 2.21 shows that Eq. (2.9) fits nicely the 2000 data.

The inset of Fig. 2.21 shows the time evolution of the Gini coefficient, as computed by the IRS (Strudler *et al.* 2003). The values of G shown in the inset of Fig. 2.21 are quite close to the theoretical value $G = 1/2$, for a purely exponential

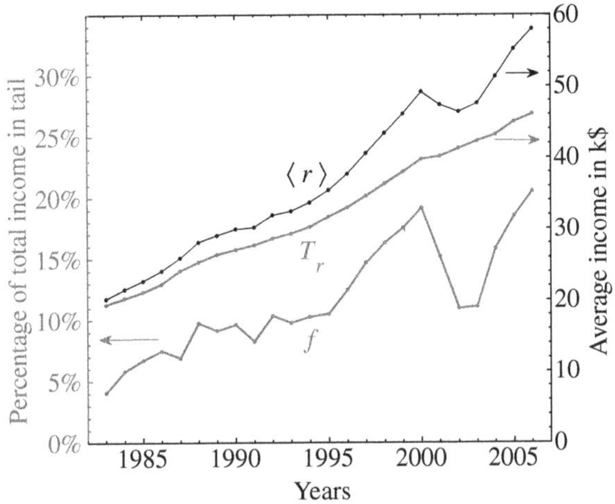

Figure 2.22 Plot of the historical evolution of the parameters $\langle r \rangle$, T_r and the fraction of income f going to the upper tail, as defined in Eq. (2.10). Reproduced from Yakovenko and Barkley Rosser (2009).

distribution. If one takes into account the upper tail using Eq. (2.9), the modified formula for the Gini coefficient becomes $G = (1 + f)/2$ (Silva and Yakovenko 2005). The inset in Fig. 2.21 shows that this new formula fits well to the IRS data for the 1990s, using the values of f deduced from Fig. 2.10. However, the values $G < 1/2$ in the 1980s could not be captured by this formula, because the Lorenz data points were slightly above the theoretical curve for 1983. Silva and Yakovenko (2005) observed that income inequality had been increasing for the past 20 years, owing to swelling of the Pareto tail, but decreased in 2001 after the stock market crash.

One can deduce that the parameter f in Eq. (2.9) and in Fig. 2.21 is given by

$$f = \frac{\langle r \rangle - T_r}{\langle r \rangle}, \tag{2.10}$$

where $\langle r \rangle$ is the average income of the whole population, and T_r is the average income in the exponential part of the distribution. Then Eq. (2.10) suitably gives a measure of the deviation of the actual income distribution from the exponential one, or an indicator of the fatness of the upper tail. Figure 2.22 shows historical evolution of the parameters $\langle r \rangle$, T_r and f given by Eq. (2.10). From the results, Silva and Yakovenko concluded that the speculative bubbles greatly increased the fraction of income going to the upper tail, but did not change income distribution of the lower class; when the bubbles inevitably collapsed, income inequality reduced.

In the previous section, we mentioned work (Clementi *et al.* 2007, 2008) where the authors had proposed the κ-generalized function to fit the income distribution,

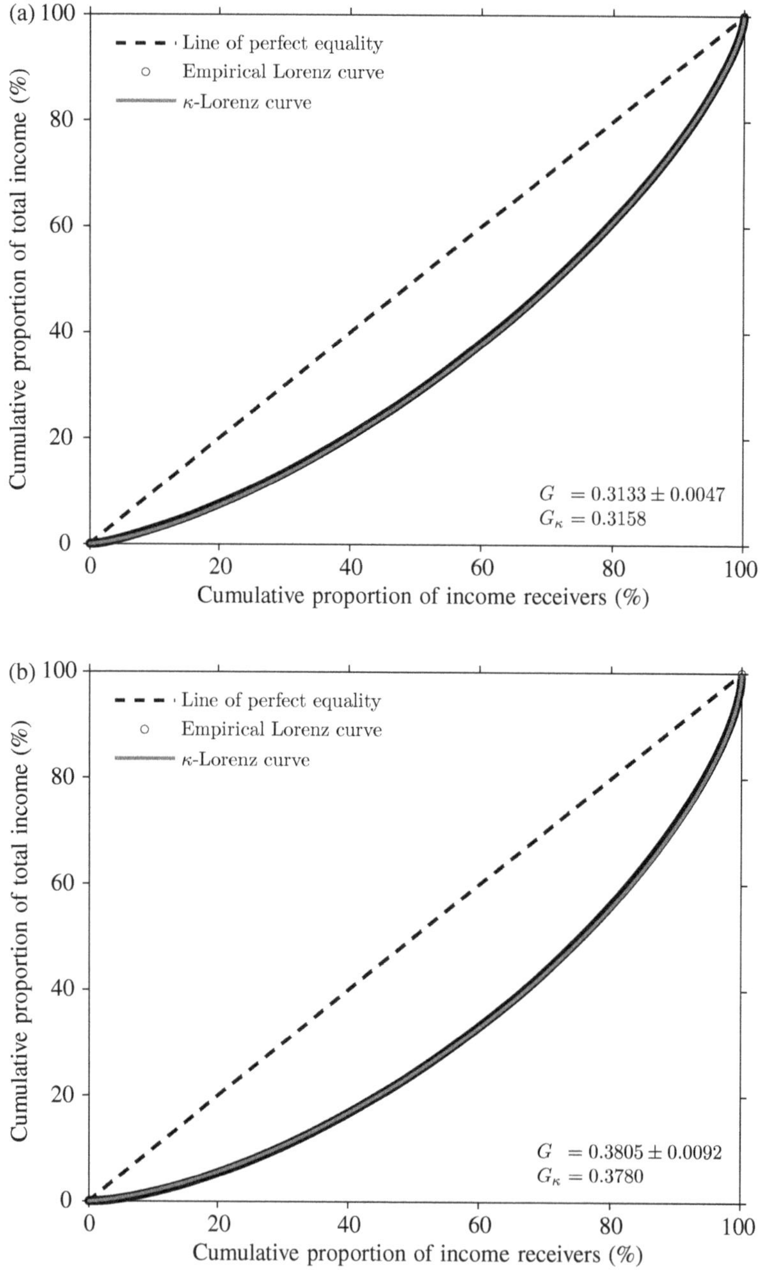

Figure 2.23 Plot of the Lorenz curve. (a) The Australian personal income distribution in 2002–3 measured in current year AUD. The US personal income distribution in 2003, measured in current year USD. In the plots, the open circles represent the empirical data points and the solid lines the theoretical curves. The dashed lines correspond to the Lorenz curves of a society in which everybody receives the same income, and thus serve as a benchmark case against which actual income distribution may be measured. Reproduced from Clementi *et al.* (2008).

Table 2.2 *Gini indices of some countries*

Country	Gini index (%)
Denmark	24.7
Japan	24.9
Sweden	25.0
Norway	25.8
Germany	28.3
India	32.5
France	32.7
Australia	35.2
UK	36.0
USA	40.8
Hong Kong	43.4
China	44.7
Russia	45.6
Mexico	54.6
Chile	57.1
Brazil	59.1
South Africa	59.3
Botswana	63.0
Namibia	70.7

From United Nations Development Programme (2004, pp. 50–53; more recent data are also available from their website).

given by Eq. (2.4):

$$p(x) = \frac{\alpha\beta x^{\alpha-1}\exp_{\kappa}(-\beta x^{\alpha})}{\sqrt{1+\kappa^2\beta^2 x^{2\alpha}}}.$$

One can derive the Lorenz curve for the κ-generalized distribution as:

$$\begin{aligned}L_{\kappa}(u) = 1 - &\frac{1+\frac{\kappa}{\alpha}}{2\Gamma\left(\frac{1}{\alpha}\right)} \frac{\Gamma\left(\frac{1}{2\kappa}+\frac{1}{2\alpha}\right)}{\Gamma\left(\frac{1}{2\kappa}-\frac{1}{2\alpha}\right)} \\ &\times \left\{ 2\alpha(2\kappa)^{\frac{1}{\alpha}}(1-u)\left[\log_{\kappa}\left(\frac{1}{1-u}\right)\right]^{\frac{1}{\alpha}} \right. \\ &\left. + B_X\left(\frac{1}{2\kappa}-\frac{1}{2\alpha},\frac{1}{\alpha}\right) + B_X\left(\frac{1}{2\kappa}-\frac{1}{2\alpha}+1,\frac{1}{\alpha}\right)\right\}, \end{aligned} \quad (2.11)$$

where $B_x(s, r)$ is the incomplete beta function given by

$$B_x(s,r) = \int_0^x t^{s-1}(1-t)^{r-1}\,dt$$

with $X = (1-u)^{2\kappa}$. Also, the Gini coefficient for the κ-generalized distribution is:

$$G_\kappa = 1 - \frac{2\alpha + 2\kappa}{2\alpha + \kappa} \frac{\Gamma\left(\frac{1}{\kappa} - \frac{1}{2\alpha}\right)}{\Gamma\left(\frac{1}{\kappa} + \frac{1}{2\alpha}\right)} \frac{\Gamma\left(\frac{1}{2\kappa} + \frac{1}{2\alpha}\right)}{\Gamma\left(\frac{1}{2\kappa} - \frac{1}{2\alpha}\right)}. \tag{2.12}$$

Using the same estimation methods and analyses as before, they obtained plots of the Lorenz curve, for the Australian personal income distribution in 2002–3 measured in current year AUD, and the US personal income distribution in 2003, measured in current year USD, as shown in Fig. 2.23.

Finally, one should mention that there have been several studies done all over the world that measure the Gini coefficients. The underlying household surveys differ in methods and in the type of data collected, and hence the distribution data are not strictly comparable across countries. However, some values of G for some countries are listed in Table 2.2. A complete list can be retrieved from Wikipedia (http://en.wikipedia.org/wiki/List_of_countries_by_income_equality).

In this chapter, we have thus seen that the income (and wealth) distributions have a sort of regularity: the top 5–10% of the very rich population in any country constitute the power law regime (Pareto tail) of the income distribution, and the rest of the population gives rise to the gamma function (or log-normal) form. In the following chapters, we shall study the socioeconomic models which have been used to reproduce such empirical characteristics.

3

Major socioeconomic modelling

In this chapter, we discuss the background and motivation for various modelling efforts carried out over the years. While some models take into account the various details as in real economic transactions, trying to fit the detailed characteristics of empirical data, others reproduce the empirical features qualitatively using very simple ideas of physical origin.

3.1 Models of income distribution

Pareto's extensive studies in Europe showed that the tail of the wealth distribution for the richer sections of society follows a power law (Pareto 1897), now known as the Pareto law. Gibrat (1931) separately worked on the same problem and proposed a law of proportionate development. Much later, Champernowne also considered this problem systematically and proposed a probabilistic theory to justify Pareto's claim (Champernowne 1953; Champernowne and Cowell 1998).

Gibrat (1931) observed that the power law distribution did not fit all the income data and hence proposed a law of proportionate effect, which states that a small change in a quantity is independent of the quantity itself. Thus the distribution of a quantity $dz = dx/x$ should be Gaussian, and hence x is log-normal. The asymptotics of this can lead to the Pareto power law. Gibrat found that the small- and middle-income ranges appear to be well described by a log-normal distribution (see Eq. (2.3)). The details of Gibrat's law and its derivation will be presented in Chapter 7.

Champernowne (1953) proposed a multiplicative model, but assumed that there is some minimum income m. For the first range, one considers incomes between m and γm, for some $\gamma > 1$; for the second range, one considers incomes between γm and $\gamma^2 m$. A person is assumed to be in class j for $j \geq 1$ if their income is between $m\gamma^{j-1}$ and $m\gamma^j$. Champernowne assumed that over each time step, the probability of an individual moving from class i to class j, denoted by p_{ij}, depends only on the value of $|j - i|$. The equilibrium distribution of individuals among classes under this assumption gives the Pareto law. This is again presented in detail in Chapter 7.

In the context of a general multiplicative model (Mitzenmacher 2004), Champernowne's model can be thought to be a special case. Moving from one class to another can be thought of as either doubling or halving the income over one time step: say, if m_t is the income after time t, then $m_t = f_t m_{t-1}$, where $f_t = 1/2$ with probability $2/3$ and $f_t = 2$ with probability $1/3$. The resultant distribution is log-normal. As long as there is a bounded minimum that acts as a lower reflective barrier to the multiplicative model, it will yield a power law instead of a log-normal distribution. The theory of the above is now well established (Gabaix 1999). Chapter 7 deals with more models and issues related to generation of income, inequality and development.

3.2 Models of wealth distribution

This section will be devoted to a detailed description of various popular and important models that address the issue of wealth distribution in a society, starting from the chemical kinetics-inspired Lotka–Volterra models, the directed polymer-inspired Bouchaud–Mézard models, a model of an evolving society and finally some models that incorporate risk aversion in investment.

3.2.1 Lotka–Volterra models

The generalized Lotka-Volterra model for wealth distribution (Levy and Solomon 1997; Richmond and Solomon 2001; Solomon and Richmond 2002) is based on the redistribution of wealth of a total number N of agents in a society. The equation

$$m_{i,t+1} = (1 + \xi_t)m_{i,t} + \frac{a}{N}\sum_j m_{j,t} - c\sum_j m_{i,t}m_{j,t} \tag{3.1}$$

combines a multiplicative random process with an autocatalytic process. Here, $m_{i,t}$ represents the amount of money assigned to an agent i at time t. The random numbers ξ_t are chosen from a positive set which has a variance V. The second term on the right-hand side of Eq. (3.1) redistributes at each time step a fraction of the total money to ensure that the money possessed by any agent is never zero as a result of this random process, which is supposed to simulate the effect of a tax or some kind of social security policy. The parameter c in the last term on the right-hand side controls the overall growth of the system, and represents external limiting factors such as the finite amount of resources and money in the economy, technological inventions, wars and disasters. It also includes internal market effects: competition between investors, adverse influence of bids on prices (e.g. when large investors sell assets to realize their profits and thereby cause prices/profits to fall). This term has the effect of limiting the growth of the total amount of money m_t in the system to values sustainable for the current conditions and resources. Clearly the equation has no stationary solution and the total money within this system can change with time.

Here the relative value of money possessed by an agent, $x_{i,t} = m_{i,t}/\langle m_t \rangle$, is independent of c. The distribution function for the relative value of money $x_{i,t}$ depends only on a and the variance V of the random variables and, if the ratio a/V is constant, the dynamics of the relative money are independent of time. Consequently, even in a non-stationary system, after some time, the relative wealth distribution will eventually converge to a time-invariant form (Levy and Solomon 1997; Richmond and Solomon 2001; Solomon and Richmond 2002).

A mean-field approximation enables one to obtain a stationary distribution function $P(x)$ of the form:

$$P(x) = \frac{\exp\left[-(\nu - 1)/x\right]}{x^{1+\nu}}. \tag{3.2}$$

The positive exponent, ν, is a ratio of parameters of the model that are related to the redistribution process and volatility of the random process. For large values of income, x, this expression indeed exhibits a Pareto-like behaviour. Numerical simulations by Malcai *et al.* (2002) of the complete dynamics of this model show this solution to be essentially exact.

A particular form of an agent exchange model was proposed by Di Matteo *et al.* (2004). In this model agents exchange money according to the following rule:

$$m_{k,t+1} = m_{k,t} + A_{k,t} + B_k m_{k,t} \sum_{j \neq k} m_{j,t}\, T(j,k|t) - m_{k,t} \sum_{j \neq k} T(k,j|t), \tag{3.3}$$

where $A_{k,t}$ and B_k are additive and multiplicative noises, respectively, the stochastic terms depicting market and social fluctuations. The term with coefficient $T(j,k|t)$ simply denotes the wealth agent k receives from j while the other term with $T(k,j|t)$ denotes the wealth agent j receives from k. Numerical calculations show that the solution of this equation yields a one-agent distribution function that agrees qualitatively with empirical income data of Australia for both the low- and high-income regions.

3.2.2 The Bouchaud–Mézard and related models

Bouchaud and Mézard (2000) discussed the appearance of Pareto tails on the basis of a very general model for the growth and redistribution of wealth. They used the physics of directed polymers and translated it to the economical framework. In this model the wealth w_i of an agent i ($i = 1, 2, \ldots, N$) is assumed to have a dynamics governed by the following set of stochastic equations:

$$\frac{dw_i(t)}{dt} = \eta_i(t)w_i(t) + \sum_{j \neq i} J_{ij} w_j(t) - \sum_{j \neq i} J_{ji} w_i(t), \tag{3.4}$$

where $\eta_i(t)$ are independent Gaussian variables of mean μ and variance $2\sigma^2$, with $\eta_i(t)w_i(t)$ being the Gaussian multiplicative process simulating the investment dynamics, and J_{ij} is the linear exchange rate between agents i and j.

The above model can be easily studied within the Fokker–Planck approach in the mean-field limit. Under mean-field approximation, one has $J_{ij} = J/N$, which simplifies the above to

$$\frac{dw_i(t)}{dt} = \eta_i(t)w_i(t) + J(\bar{w} - w_i), \tag{3.5}$$

where $\bar{w} = \sum_i w_i/N$ is the mean wealth. It can be easily shown that the average wealth increases in time as $\bar{w}_t = \bar{w}_0 \exp[(\bar{\eta} + \sigma^2)t]$. Writing everything in terms of the relative wealth $\tilde{w}_i = w_i/\bar{w}_t$, Eq. (3.5) becomes

$$\frac{d\tilde{w}_i}{dt} = [\eta_i(t) - \bar{\eta} - \sigma^2]\tilde{w}_i + J(1 - \tilde{w}_i). \tag{3.6}$$

The probability distribution $P(\tilde{w}, t)$ for the stochastic differential equation (3.6) is governed by the Fokker–Planck equation

$$\frac{\partial P}{\partial t} = \frac{\partial[J(\tilde{w} - 1) + \sigma^2\tilde{w}]P}{\partial \tilde{w}} + \sigma^2 \frac{\partial}{\partial \tilde{w}}\left(\tilde{w}\frac{\partial(\tilde{w}P)}{\partial \tilde{w}}\right). \tag{3.7}$$

The stationary solution ($\partial P/\partial t = 0$) for the above is analytically found to be

$$P(\tilde{w}) = A\frac{\exp[(1 - \nu)/\tilde{w}]}{\tilde{w}^{1+\nu}}, \tag{3.8}$$

where $\nu = 1 + J/\sigma^2$, and $A = (\nu - 1)^\nu/\Gamma[\nu]$, In the large w limit, the tail of the distribution shows a power law behaviour with Pareto exponent ν. One of the most interesting results of such a model is the existence of a phase transition, separating a phase in which the total wealth of a very large population is concentrated in the hands of a few individuals (corresponding, to the case $\nu < 1$) from a phase in which it is shared by a finite fraction of the population. An interesting outcome is that wealth tends to be very widely distributed with limited exchanges, either in amplitude or topologically.

Scafetta *et al.* (2004) discussed a few interesting consequences of the above elegant model. The main aspect of the mean-field model was that the agents exchange the same percentage of wealth they have, meaning that the poorer agent receives an unrealistic amount of wealth from the rich. Also, in the absence of the multiplicative process, Eq. (3.5) has a solution

$$w_i(t) = \bar{w} + (w_i(0) - \bar{w})\exp[-Jt], \tag{3.9}$$

which implies that the wealth of all agents converges asymptotically to the mean wealth $\bar{w}$, an effect of 'equalizing' the society. Therefore, in this framework, there has to be investments which will create the economic inequality. They proposed a modification of the Bouchaud–Mézard model:

$$w_i(t+1) = w_i(t) + r_i\xi(t)w_i(t) + \sum_{j\neq i} W_{ij}(t), \tag{3.10}$$

where the investment term is rewritten as $r_i\xi(t)w_i(t)$; r_i, being the standard deviation of the Gaussian variable $r_i\xi(t)$, is now called the *individual investment index*; W_{ij} is the actual amount of wealth that is exchanged between i and j, and hence changes in each transaction. The justification for the term W_{ij} is as follows: any trade consists of an exchange of an asset and money, flowing in directions opposite to each other. Thus, the authors define W_{ij} = value − price, with a non-zero transfer of wealth only if the 'value' and 'price' differ. Thus $W_{ij} = -W_{ji}$, which also ensures the conservation of wealth in a particular trade. For simplicity, W_{ij} are assumed to be Gaussian random variables with probability density

$$p(W_{ij}) = \frac{1}{\sigma\sqrt{2\pi}}\exp\left[-\frac{(W_{ij} - \bar{W}_{ij})^2}{2\sigma^2}\right], \tag{3.11}$$

where $\bar{W}_{ij}$ is the mean wealth that can be exchanged between i and j and $\sigma = h\omega_{ij}$ is the standard deviation of the distribution of W_{ij} and assumed to depend on $\omega_{ij} = \omega_{ji} = \min[w_i, w_j]$. $0 \leq h \leq 1$ ensures that the fluctuation of the wealth is a proper fraction of the wealth of the poorer trader, and could be interpreted as the *poverty index*.

3.2.2.1 The Bouchaud–Mézard model on different networks

Garlaschelli and Loffredo (2008) considered various interaction topologies for the Bouchaud–Mézard model. As in standard literature they specified the topology by an *adjacency matrix* a_{ij}, whose elements are $a_{ij} = 1$ if there is a link from i to j, and $a_{ij} = 0$ otherwise. $a_{ij} = a_{ji}$ for undirected networks. The basic topological characteristic of each vertex is given by its *degree* $k_i = \sum_j a_{ij}$, and the statistical distribution of the degrees of all vertices is given by $\Pi(k)$. Interactions take place only between connected agents and hence $J_{ij} = (J/N)a_{ij}$. In the two extreme scenarios (1) when agents are independent, one can treat $J_{ij} = 0$ for each pair i, j, corresponding to a network with no edges, which will yield a log-normal distribution (2.3); (2) when all agents are connected to one another as in the Bouchaud–Mézard model, $J_{ij} = J/N$ for each i, j pair, which yields a distribution with a power law tail (3.8). Thus, it was clear that one should expect something different between these two extreme topologies. In case of the Erdös–Renyi random graph, no mixed form for $P(w)$ is observed between the two extremes of $p = 0$ where Gibrat's law holds and $p = 1$ where $P(w)$ is a power law. However, the case is different for a *small world* network. Starting from a regular d-dimensional lattice in which each vertex is connected to the first r neighbours, one rewires the network randomly with probability p. Thus, $p = 0$ is the original regular network, while $p = 1$ is a fully rewired random network. For the $d = 1$ case, it has been seen that the form of $P(w)$ depends on vertex degree $k = 2r$. Log-normal distributions are found for small k while Pareto tails appear for large k. However, the interesting aspect lies in the region of intermediate p and $\langle k \rangle$ where the $P(w)$ fits well to a combined distribution of log-normal for small w and power law for large w. On the Barabasi–Albert model for scale-free networks, the Bouchaud–Mézard model has log-normal and power law wealth distributions, respectively, for small and large values of average degree $\langle k \rangle$, with the intermediate region showing no mixed form (Fig. 3.1).

In order to address the scenario using simple heterogeneous graphs, the authors consider an undirected network of N vertices, out of which M are fully connected and the remaining $N - M$ are isolated ($k_i = M - 1$ for $i = 1, \ldots, M$ and $k_i = 0$ for $i = M + 1, \ldots, N$). Thus, owing to the evolution (3.4), one gets a mixed form of the wealth distribution $P(w)$ given by

$$P(w) = \frac{M}{N} P_1(w) + \left(1 - \frac{M}{N}\right) P_2(w), \tag{3.12}$$

where $P_1(w)$ is the power law distribution and $P_2(w)$ is a log-normal, with a single control parameter M/N.

Inferring from previous results on the regular graph, one can replace the isolated nodes by a regular chain of $N - M$ nodes. Another alternative is to consider a dense core of M nodes and each of the rest N_M with one connection to the core. All of these yield the same type of mixed distributions. The inference drawn from such exercises is that one requires a network with structural heterogeneities, with coexisting regions of dense cores and periphery links of low density.

3.2.3 Models for evolving society

Ideas from network science have been applied to explain non-linear growth, even in wealth distribution (Dorogovtsev and Mendes 2003a). The growth is modelled by assuming an input flow of capital proportional to t^{α}, which means that $\alpha = 0$

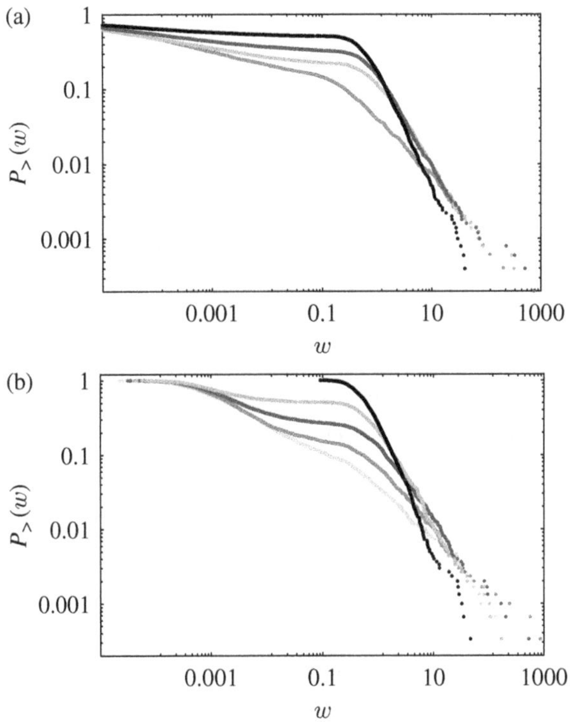

Figure 3.1 Cumulative wealth distribution $P_>(w)$ for the Bouchaud–Mézard model on a mixed network for different values of M/N: from top to bottom, $M/N = 1/2, 1/4, 1/8, 1/16$. In all cases $N = 5000$, $J = \sigma^2 = 0.05$ and $m = 1$. The wealth is rescaled to its average. (b) Same for the Bouchaud–Mézard model on an octopus network for the same sets of values of M/N. In all cases $N = 3000$, $J = \sigma^2 = 0.05$ and $m = 1$. The wealth is rescaled to its average. Reproduced from Garlaschelli and Loffredo (2008).

corresponds to a stable society while positive and negative α models correspond to evolving and degrading societies, respectively.

The model is based on the idea that money attracts money. The additional attractiveness A is proportional to the average wealth, and may be provided at birth, $A(s)$ ($s < t$ is the time of birth of the individual).

3.2.3.1 *Stable society*

Starting from an initial capital m_s, let m extra wealth be distributed at each time step. The total flow of wealth is thus $m + m_s$. The additional attractiveness A is assumed to be constant. A fraction p of the flow m is distributed among all individuals randomly and $(1 - p)m$ preferentially with probability proportional to the wealth. Thus the continuum approach equation for the average wealth of an individual, $\bar{w}(s,t)$, is

$$\frac{\partial \bar{w}(s,t)}{\partial t} = \frac{pm}{t} + (1-p)m \frac{\bar{w}(s,t) + A}{\int_0^t du[\bar{w}(u,t) + A]}, \tag{3.13}$$

subject to initial condition $\bar{w}(0,0) = 0$ and boundary condition $\bar{w}(t,t) = m_s$. Integrating the above equation yields: $\int_0^t ds \bar{w}(s,t) = (m + m_s)t$, and subsequent calculations produce a power law distribution of wealth with Pareto exponent

$$\nu = 1 + \frac{pm + m_s + A}{(1-p)m}. \tag{3.14}$$

Here, $\nu > 1$ always. This is the case for stable societies, where the average wealth per individual does not change in time.

3.2.3.2 Developing and degrading society

Here, the starting capital is assumed to be proportional to the average wealth of the society at the time of the individual's birth: $m_s(t) = bmt^\alpha$, b being a positive constant. Out of the wealth mt^α distributed among the individuals at each time, a fraction p is distributed equally, while $(1-p)$ fraction is distributed preferentially (as in the previous case). Without loss of generality, it is assumed that $A(s,t) = 0$. Then

$$\frac{\partial \bar{w}(s,t)}{\partial t} = mt^\alpha \frac{p}{t} + (1-p)mt^\alpha \frac{\bar{w}(s,t)}{\int_0^t du\, \bar{w}(u,t)}. \tag{3.15}$$

The initial and boundary conditions are $\bar{w}(0,0)$ and $\bar{w}(t,t) = bmt^\alpha$. Integration yields $\int_0^t ds\, \bar{w}(s,t) = m(1+b)t^{\alpha+1}/(\alpha+1)$. Equation (3.15) at various values of the parameters p, b and α describes different societies:

(1) When $\alpha > (1-p)/(p+b)$, i.e. $p > (1-b\alpha)/(1+\alpha)$, the wealth distribution is exponential. This is referred to as a 'super-fair' society.
(2) When $\alpha < (1-p)/(p+b)$, the wealth distribution is a power law with Pareto exponent

$$\nu = 1 + \frac{(1+\alpha)(p+b)}{1-p-\alpha(p+b)}. \tag{3.16}$$

Observe that $\nu = 1$ at $\alpha = -1$, and this corresponds to a wealth condensation transition from a fair society ($\nu > 1$ for $\alpha > -1$) to an unfair one ($\nu < 1$ for $\alpha < -1$; few individuals keep a finite fraction of the total wealth). It is also important to note that the position of the condensation transition does not depend on the particular values of p and b (Fig. 3.2).

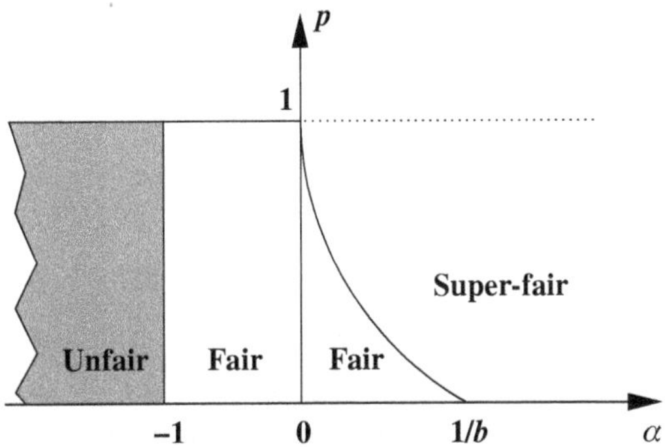

Figure 3.2 Phase diagram of the evolving society model. The wealth distribution in the super-fair society is exponential. The wealth distribution has a power law in the 'fair' society ($\nu > 1$) as well as in the 'unfair' society ($\nu < 1$).

3.2.4 Models of wealth distribution with a risk aversion factor

In a competing market, agents will always try to improve their economic situation. The poorest agent feels the most of this pressure to rise up the economic ladder and will try to undertake certain measures such as a change in the trade strategy or production methods or borrow money. Since there is an uncertainty involved in the outcome, it can be modelled as a random change in the wealth parameter of the agent. Pianegonda *et al.* (2003) proposed a process in which one considers agents to lie on a ring with wealth exchange possible only with the nearest neighbours. Starting from an initial random distribution of wealth (between 0 and 1), the poorest agent is allowed to change its state randomly, at the expense of the neighbours, i.e. the neighbours equally divide the wealth gained or lost by the particular agent. Given that there is no restriction on the amount of wealth, negative wealth (or debt) is thus allowed in this model. The stationary wealth distribution resulting from such an extremal dynamics is one where $P(w) = 0$ for $w < w_T$ while there is an exponentially decaying $P(w)$ following it. For 1 – d, they found $w_T \approx 0.4$. The globally coupled (mean field) model also shows similar features, albeit a lower value of w_T. The computed Gini indices fall in a regime close to most economies.

In another model further developed by the group (Iglesias *et al.* 2004), they introduce the concept of a *risk aversion factor*. Agents are endowed with a risk aversion factor β_i such that $1 - \beta_i$ is the percentage of wealth that an agent is willing to risk, meaning that $\beta_i = 0$ is a radical agent risking its all assets, while $\beta_i = 1$ is a totally conservative agent not willing to risk at all. It is prescribed that no agent can win more than what is invested, so that the amount that will be exchanged is the minimum of the available resources of each agent $dw = \min[(1 - \beta_1)w_1; (1 - \beta_2)w_2]$. Finally, it is argued that a stable society requires the poor to have an advantage in transactions with the wealthy and are protected by particular rights and marketing freedom. Two cases were considered: (1) the poor are favoured with a probability p between 0.5 and 1 and (2) a random value of p given by

$$p = \frac{1}{2} + f \times \frac{w_1 - w_2}{w_1 + w_2}, \tag{3.17}$$

w_1 being the wealth of the richer partner, and f is a factor between 0 and 0.5. Two different cases were considered: (A) when β and p are both uniform and (B) when β and p are both random. For the case A, the typical wealth distribution features could be broadly distinguished into regions in the β–p phase space. Wealth distributions in region I are narrow and Gaussian-like, which the authors called *utopian socialism* because almost all agents have the same income with a small dispersion. Region II has Gaussian-like distributions but skewed to higher values of wealth; therefore, they named it *liberal socialism*. Region III has mixed wealth distribution: Gaussian-like for low wealth values and exponential for high wealth values, and they called it *moderated capitalism*. In the last region (region IV) the wealth distributions are exponentials with a tendency to power laws, so they called this region *ruthless capitalism* (Fig. 3.3). Beyond this region, only one or two agents hold all the resources, leaving the rest without anything.

In the same model with extremal dynamics, the dynamics at low p phase freezes since the agent with minimum wealth has no asset to trade with. However, for $0.7 \leq p \leq 1$, an exponential distribution of wealth is produced, with most agents consisting of the middle class. For $\beta \geq 0.7$ and $p \approx 1$, the middle class is split into two income regions separated by a gap.

For case B, that is, when both β and p are random, the risk aversion parameter is taken to be disordered, i.e. β_i is randomly drawn from a uniform distribution in (0, 1). This is a quenched disorder, so that each agent maintains its value of β_i throughout. For the case $f = 0$, a trade not favouring either of the partners, the wealth distribution steadily converges to a delta function at $w = 0$, most agents having nothing while almost the entire wealth is held by one or two agents. In this case, since each agent risks a part of its wealth at each exchange, and while there is no restriction on the amount it can lose in successive trades, this is nothing but a multiplicative process with an attractor at $w = 0$. On the other hand, when there is asymmetry, it favours the poorer agent on average. The result is a distribution with a region for the poor, followed by a peak and then a middle class decaying as a power law $P(w) \sim w^{-\alpha}$, with α depending on f, with $\alpha \approx 2$ for $f = 0.5$ (Fig. 3.4). It is important to note the behaviour of the risk–wealth correlation plot: the higher values of income correspond to high risk aversion while the highest individual wealths correspond to risk-loving agents, and it is no surprise to find that the lowest incomes correspond to risky agents.

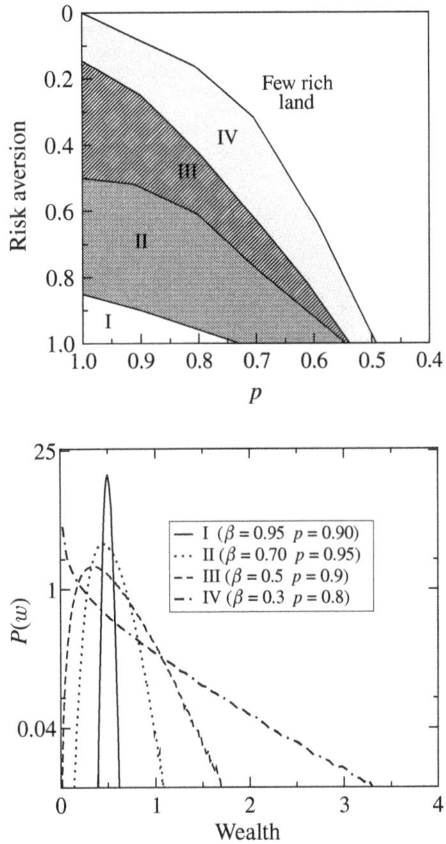

Figure 3.3 Model with uniform β and p, for $N = 10^5$ with 10^3 transactions per agent. Region I corresponds to a very narrow wealth distribution, utopian socialism; regions II and III present skewed Gaussians; and region IV corresponds to an exponential distribution. Outside these regions there is no true wealth distribution because in the 'few rich land' only a few agents have all the available resources while the others have strictly zero wealth. Reproduced from Iglesias *et al.* (2004).

Performing extremal dynamics on such a model, one finds a *poverty line* below which there are a few people, while the rest of the distribution follows an exponential behaviour $P(w) \propto \exp[-a(w - w_0)^2]$ with $a \approx 1.1$ and $w_0 \approx 0.7$ for $f = 0.5$.

In another work, Fuentes *et al.* (2006) considers the case when the partners in the transaction have previous knowledge of the winning probability and adjust their risk aversion taking this into consideration. The results indicate that a relatively egalitarian society is obtained when the agents' risk is directly proportional to their winning probabilities. However, a case contrary to this produces a wealth distribution and Gini indices that resemble empirical data. The authors conclude that this indicates that, at least for their very simple model, either agents have no

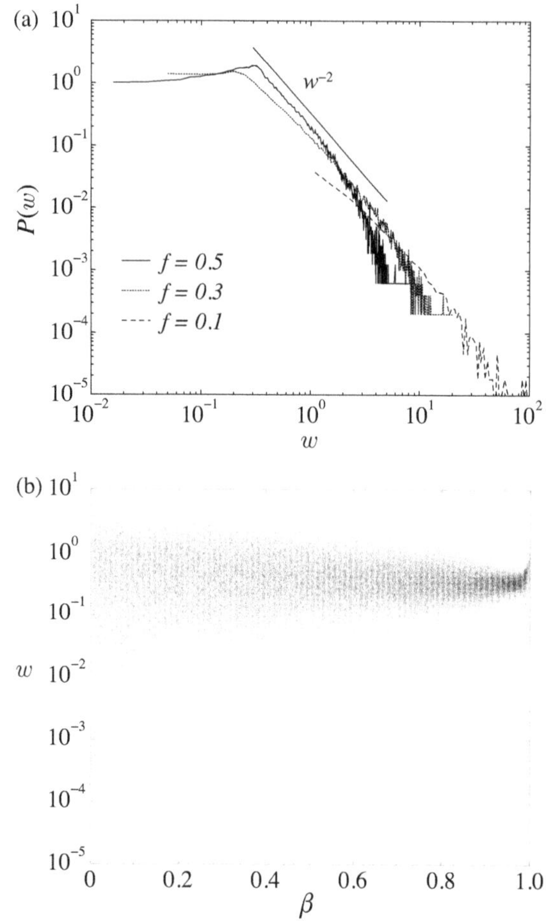

Figure 3.4 (a) Wealth distribution for random β, the distribution is calculated for $N = 10^5$ and 10^4 exchanges per agent on average. Results are shown for three values of the asymmetry parameter. The Ax^2 curve is also shown as a guide. (b) Correlation between wealth and saving parameter for $N = 10^5$, $f = 0.5$. Reproduced from Iglesias *et al.* (2004).

knowledge of their winning probabilities or they exhibit an 'irrational' behaviour risking more than is reasonable.

3.2.5 Asset exchange models

There is a whole body of literature that considers mainly conservative wealth models, in which wealth is transferred or shared from one agent to the other, and it is appropriate to call them 'asset exchange models' to be consistent with the existing literature. Ispolatov *et al.* (1998) wrote an interesting article that describes a very simple framework under which 'asset exchange' models can be generalized into additive and multiplicative models, and we discuss these in detail in the following.

The basic structure of the model is as follows: there is a population of traders who possess some 'asset', probably bearing the same meaning as 'wealth' as we prefer to use in other places in this book. It is assumed that the asset is quantized in some basic unit, such that individuals hold integer amount of the same, and the wealth of individuals evolves according to some prescribed rules of interaction between random pairs of individuals.

3.2.5.1 Additive asset exchange

Here, one considers the case in which one unit of asset is transferred between a random pair of individuals. In the case in which an individual has no asset, it is 'bankrupt' and can no longer participate in the trading process. The authors consider three different realizations of the additive rule: the 'random', 'greedy' and 'very greedy' types.

In the 'random' additive asset exchange, the direction of the asset transfer is independent of the assets of the traders. One unit of asset is exchanged between trading partners, represented in the reaction scheme as $(j, k) \to (j \pm 1, k \mp 1)$. If $n_k(t)$ is the density of individuals with capital k, under a mean-field description, $n_k(t)$ evolves as

$$\frac{dn_k(t)}{dt} = N(t)\left[n_{k+1}(t) + n_{k-1}(t) - 2n_k(t)\right], \tag{3.18}$$

with $N(t) \equiv M_0(t) = \sum_{k=1}^{\infty} n_k(t)$ as the population density. The first two terms show the gain in $n_k(t)$ owing to the interactions $(j, k+1) \to (j+1, k)$ and $(j, k-1) \to (j-1, k)$, respectively, and the last term shows the loss in $n_k(t)$ owing to the interactions $(j, k) \to (j \pm 1, k \mp 1)$. By suitably defining a modified time variable,

$$T = \int_0^t dt' N(t'), \tag{3.19}$$

Eq. 3.18 is reduced to a discrete differential equation

$$\frac{dn_k(T)}{dT} = n_{k+1}(T) + n_{k-1}(T) - 2n_k(T). \tag{3.20}$$

The rate equation for the poorest density looks slightly different, $dn_1/dT = n_2 - 2n_1$, but could be written in the same form as Eq. (3.20) on imposing the boundary condition $n_0(T) = 0$. Equation (3.20) can be solved for arbitrary initial conditions. For example, if everyone starts with unit asset $n_k(0) = \delta_{k1}$, one eventually gets

$$N(t) \simeq \left(\frac{2}{3\pi t}\right)^{1/3} \tag{3.21}$$

and

$$n_k(t) \simeq \frac{k}{3t} \exp\left[-\left(\frac{\pi}{144}\right)^{1/3} \frac{k^2}{t^{2/3}}\right]. \tag{3.22}$$

It is interesting to note that the last equation can be expressed as $n_k(t) \propto N^2 x\, e^{-x^2}$, with the scaling variable $x \propto kN$. This scaling solution is also the basin of attraction for almost all exact solutions. Ispolatov *et al.* (1998) claim that, for any initial condition with $n_k(0)$ decaying faster than k^{-2}, the system reaches the scaling limit $n_k(t) \propto N^2 x\, e^{-x^2}$. However, an initial condition $n_k(0) \sim k^{-1-\alpha}$, with $0 < \alpha < 1$, converges to an alternative scaling limit which depends on α (e.g. as in Derrida *et al.* 1991). These solutions follow a slower decay of the total density, $N \sim t^{-\alpha/(1+\alpha)}$, while the scaling form of the wealth distribution is

$$n_k(t) \sim N^{2/\alpha} \mathcal{C}_\alpha(x), \quad x \propto kN^{1/\alpha}, \tag{3.23}$$

with the scaling function

$$\mathcal{C}_\alpha(x) = e^{-x^2} \int_0^\infty du\, \frac{e^{-u^2} \sinh(2ux)}{u^{1+\alpha}}. \tag{3.24}$$

A Laplace transform will produce an asymptotic distribution of $x^{-1-\alpha}$ as in the initial condition. The authors conclude that this anomalous scaling in the solution to the diffusion equation is a direct consequence of the extended initial condition.

In another variant, called the 'greedy' exchange, the richer person takes one unit of capital from the poorer person, represented by the reaction scheme $(j, k) \to (j+1, k-1)$ for $j \geq k$. Under the rate equation approximation, the densities $n_k(t)$ evolve according to

$$\frac{dn_k}{dt} = n_{k-1} \sum_{j=1}^{k-1} n_j + n_{k+1} \sum_{j=k+1}^{\infty} n_j - n_k N - n_k^2. \tag{3.25}$$

The first two terms represent the gain in $n_k(t)$ owing to the interaction between pairs of individuals of capitals $(j, k-1)$, with $j < k$ and $(j, k+1)$ with $j > k$, respectively, while the last two terms account for the corresponding losses of $n_k(t)$. It is easy to see that the wealth density $M_1 \equiv \sum_{k=1}^{\infty} k\, n_k(t)$ is conserved and that the population density obeys

$$\frac{dN}{dt} = -n_1 N. \tag{3.26}$$

Equation (3.25) are conceptually similar to the Smoluchowski equations for aggregation with a constant reaction rate (Ziff 1984), but mathematically more complex, and Eq. (3.26) is amenable to a scaling solution. It turns out that the scaled wealth distribution $\mathcal{C}(x)$ coincides with the zero-temperature Fermi distribution:

$$\mathcal{C}(x) = \begin{cases} \mathcal{C}(0), & x < x_f, \\ 0, & x \geq x_f. \end{cases} \tag{3.27}$$

Thus the scaled profile has a sharp front at $x = x_f$, with $x_f = 1/\mathcal{C}(0)$. Thus the unscaled wealth distribution $c_k(t)$ reads

$$n_k(t) = \begin{cases} 1/(2t), & k < 2\sqrt{t}, \\ 0, & k \geq 2\sqrt{t}, \end{cases} \tag{3.28}$$

and the total density is $N(t) = t^{-1/2}$. Further calculations indicated that the width of the front region behaves as $W \sim t^{1/4}$, or a relative width $w = W/k_f \sim t^{-1/4}$, so that the appropriate scaling ansatz for the front region is

$$n_k(t) = \frac{1}{t} X(\xi), \quad \xi = \frac{k - 2\sqrt{t}}{t^{1/4}}. \tag{3.29}$$

However, their analysis of data shows $w \sim t^{-\alpha}$ with $\alpha \approx 1/5$.

In the 'very greedy exchange', the events occur at a rate equal to the difference of assets of the two traders. This implies that a person is more likely to take away assets from a much poorer person rather than from someone of slightly less wealth. The corresponding rate equations are

$$\frac{dn_k}{dt} = n_{k-1} \sum_{j=1}^{k-1} (k-1-j) n_j + n_{k+1} \sum_{j=k+1}^{\infty} (j-k-1) n_j - n_k \sum_{j=1}^{\infty} |k-j| n_j, \tag{3.30}$$

while the total density obeys

$$\frac{dN}{dt} = -n_1(1-N), \tag{3.31}$$

under the assumption that the (conserved) total wealth density is set equal to 1, that is, $\sum k n_k = 1$. Ispolatov *et al.* (1998) again used the scaling argument, and found that $N \simeq [\mathcal{C}(0)t]^{-1}$ and

$$\mathcal{C}(x) = (1+\mu)(1+\mu x)^{-2-1/\mu}, \tag{3.32}$$

with $\mu = \mathcal{C}(0) - 1$. The scaling approach finds a whole class of solutions characterized by μ and additional information is required to select the kind of solutions appropriate for the models considered. As a matter of fact, the last equation has different solutions depending on the sign of μ. For $\mu > 0$, there is an extended non-universal power law distribution, while for $\mu = 0$ the solution is the pure exponential, $\mathcal{C}(x) = e^{-x}$; both of these solutions can be rejected with the argument that the wealth distribution cannot extend over an unbounded domain if the initial wealth extends over a finite range. The accessible solutions correspond to the range $-1 < \mu < 0$, where the distribution is compact and finite, with $C(x) \equiv 0$ for $x \geq x_f = -\mu^{-1}$. Eventually it can be shown that the wealth distribution is

$$n_k(t) = \begin{cases} 4/t^2, & k < t, \\ 0, & k \geq t. \end{cases} \tag{3.33}$$

Another interesting feature arises if the interaction rate is sufficiently greedy, a finite fraction of the total capital is possessed by a single individual – 'gelation' occurs. The authors identify the regime at which this happens, in terms of a parameter describing the feature of the kernel function.

3.2.5.2 *Multiplicative asset exchange*

In many economic transactions, the amount of asset transferred is a finite fraction of the initial asset of one of the participants, a simple realization of which is given by the reaction scheme $(x, y) \to (x - \alpha x, y + \alpha x)$, where $0 < \alpha < 1$ represents the fraction of the loser's asset that is gained by the winner. This process preserves the non-zero nature of the asset of any individual, although it can become vanishingly small. As in the additive processes, the authors consider the 'random' and 'greedy' variants of the multiplicative exchange process (Ispolatov *et al.* 1998).

The random multiplicative process is expressed as a rate equation

$$\begin{aligned} \frac{\partial n(x)}{\partial t} = \frac{1}{2} \int \int dy\, dz\, n(y) n(z) \times [-\delta(x - z) - \delta(x - y) \\ + \delta(y(1 - \alpha) - x) + \delta(z + \alpha y - x)], \end{aligned} \tag{3.34}$$

where the first two terms denote the loss of $n(x)$ owing to the interaction of an individual of capital x, the next term denotes the gain in $n(x)$ by the lossy interaction $(x/(1 - \alpha), y) \to (x, y + \alpha x/(1 - \alpha))$, and the last term also denotes the gain in $n(x)$ by the gaining interaction $(y, x - \alpha y) \to (y(1 - \alpha), x)$. After integrating over the delta functions, this becomes

$$\frac{\partial n(x)}{\partial t} = -n(x) + \frac{1}{2(1 - \alpha)} n\left(\frac{x}{1 - \alpha}\right) + \frac{1}{2\alpha} \int_0^x dy\, n(y)\, n\left(\frac{x - y}{\alpha}\right), \tag{3.35}$$

where the total density is set to unity. This rate equation describes a diffusive-like process on a logarithmic scale, excepting the (third) term, which describes hopping to the right and is non-local and two-body in character.

It can be easily verified that the exponential form $n(x) = Be^{-bx}$ satisfies the steady-state version of the rate equation, Eq. (3.35), if and only if $\alpha = \frac{1}{2}$ and $B = b$, which corresponds to the case that the winner receives one-half of the assets of the loser. The exact steady wealth distribution is purely exponential $n(x) = M^{-1} \exp(-x/M)$. However, for a general $0 < \alpha < 1$, the tail is again an exponential, $n(x) \simeq 2b(1 - \alpha)e^{-bx}$. Further, for $x \ll 1$, the authors claim that

$n(x) \sim x^{\lambda}$ is the asymptotic solution, with exponent $\lambda = -1 - \ln 2/\ln(1-\alpha)$. It can be observed that λ is positive when $\alpha < 1/2$, so that the density of the paupers is vanishingly small. The authors provide a heuristic justification for this phenomenon: for $\alpha < 1/2$ an unfavourable interaction leads to a relatively small loss of asset, which is more than compensated for by favourable interactions so that a poor individual could come out of its state of poverty. On the other hand, for $\alpha > 1/2$, unfavourable interactions produce a large and persistent 'poor class', with its number diverging as a power law in the limit of vanishing wealth. These features are confirmed in numerical simulations.

In the 'greedy' variant of the multiplicative exchange, the rate equation for greedy multiplicative exchange is

$$\frac{\partial n(x)}{\partial t} = -n(x) + \frac{1}{1-\alpha} n\left(\frac{x}{1-\alpha}\right) N\left(\frac{x}{1-\alpha}\right) + \frac{1}{\alpha}\int_{\frac{x}{1+\alpha}}^{x} dy\, n(y)\, n\left(\frac{x-y}{\alpha}\right), \tag{3.36}$$

where $N(x) = \int_x^{\infty} dz\, n(z)$ is the population density whose wealth exceeds x. Numerical simulations show that a continuously evolving power law wealth distribution arises, with $n(x,t) \propto 1/(xt)$, for wealth in the range $(1-\alpha)^t < x < t$ where these cut-offs correspond to the poorest and richest individuals, respectively. The authors also show that a vanishing fraction of people, of order $\ln t/t$, possess an overwhelming amount, of order $t/\ln t$, of the total wealth.

Hayes (2002) proposed two models of asset exchange in a closed, non-evolving economy based on simple exchange rules: yard-sale (YS) and theft-and-fraud (TF), as generalizations of existing models. In the YS model, the amount of wealth exchanged is a finite fraction of that of the poorer trader, and the resultant distribution corresponds to a monopoly, where all the wealth accumulates with one trader (Chakraborti 2002). In the TF model, the trading pair randomly chooses the loser, and the amount of wealth exchanged is a random fraction of the donor. Thus the rich trader has more to lose while the poor trader has more to gain. The resultant equilibrium distribution is exponential.

Some extensions of these models (Sinha 2003, 2005) produce distributions which are of considerable interest. In an asymmetric exchange model (Sinha 2005), the wealth dynamics is defined by:

$$w_i(t+1) = \begin{cases} w_i(t) + \epsilon\left(1 - \tau\left[1 - \frac{w_i(t)}{w_j(t)}\right]\right) w_j(t), & \text{if } w_i(t) \le w_j(t), \\ w_i(t) + \epsilon w_j(t), & \text{otherwise,} \end{cases} \tag{3.37}$$

ϵ is a random number between 0 and 1. Here, $\tau = 0$ corresponds to the random exchange model (Drăgulescu and Yakovenko 2000), while $\tau = 1$ corresponds to the minimum exchange model (Chakraborti 2002; Hayes 2002). In general, the relation between agents is asymmetric and the richer agent dictates the terms of trading. τ is known as the 'thrift' parameter, and it measures the degree to which the richer agent is able to use its power. If one considers a uniform distribution of τ among agents between 0 and 1, one observes a power law distribution for larger wealth, with Pareto exponent 1.5 (Sinha 2003).

We will devote the next chapter to a special class for these models which are analogous to the kinetic theory of gases.

3.3 Statistical equilibrium theory of markets

It is important at this stage to provide an outline of the statistical equilibrium theory of markets, introduced in Foley (1994) as an alternative to the conventional competitive equilibrium theory originated by Walras (1874–7) and by Marshall (1920). Conventionally, an auctioneer collects orders from buyers and sellers and determines an equilibrium price that clears the market, subject to budget constraints and utility preferences of the agents. In this new theory, all transactions occur at the same price. However, it is not clear how the real markets would actually converge to this price. Now there is a probability distribution of trades at different prices, and market clearing is achieved statistically.

Foley (1994) studied an ensemble of N economic agents trading n different types of commodities. An n-component vector $\boldsymbol{x}_j = (x_j^{(1)}, x_j^{(2)}, x_j^{(3)}, \ldots, x_j^{(n)})$ represents the state of each agent $j = 1, 2, \ldots, N$, where each component of this vector represents a possible trade that the agent j is willing to perform with a given commodity. An increase, a decrease and no change in the stock of a given commodity for the agent j are represented by positive, negative and zero values of x_j. All trades in the system are subject to the global constraint

$$\sum\nolimits_j \boldsymbol{x}_j = 0, \tag{3.38}$$

which represents the conservation of commodities during trading (only get transferred from one agent to another). An increase $x_i > 0$ of a commodity stock for an agent i must be compensated by a decrease $x_j < 0$ for another agent j, keeping the algebraic sum (3.38) of trades zero. However, Eq. (3.38) does not require a bilateral balance of transactions between pairs of agents and allows for multilateral trades.

If the number of agents doing the trade $\boldsymbol{x}_k$ is N_k, then the constraint (3.38) can be rewritten as

$$\sum\nolimits_k \boldsymbol{x}_k N_k = 0. \tag{3.39}$$

For a given set of N_k, the multiplicity gives the number of different combinations of individual agents corresponding to the trades $\boldsymbol{x}_k$, subject to N_k being fixed.

Maximizing the entropy subject to the constraints (3.39), we have

$$P(\boldsymbol{x}_k) = \frac{N_k}{N} = c\, e^{-\boldsymbol{\pi}\cdot\boldsymbol{x}_k}, \tag{3.40}$$

where c is a normalization constant, and $\boldsymbol{\pi}$ is the n-component vector of Lagrange multipliers introduced to satisfy the constraints (3.39). Foley (1994) interpreted $\boldsymbol{\pi}$ as the vector of entropic prices. The probability of an agent performing the set of trades $\boldsymbol{x}$ depends exponentially on the volume of the trades: $P(\boldsymbol{x}) \propto \exp(-\boldsymbol{\pi}\cdot\boldsymbol{x})$ (Yakovenko 2012).

Foley (1996) applied the general theory (Foley 1994) to a simple labour market. This model consists of two classes of agents – employers (firms) and employees (workers), trading in two commodities ($n = 2$): $x^{(1)} = w$ is wage, supplied by the firms and taken by the workers, and $x^{(2)} = l$ is labour, supplied by the workers and taken by the firms. For each worker, the offer set includes the line $\boldsymbol{x} = (w > w_0, -1)$, where -1 is the fixed offer of labour in exchange for any wage w greater than a minimum wage w_0. The offer set also includes the point $\boldsymbol{x} = (0, 0)$, which offers no labour and no wage, i.e. the state of unemployment. For each firm, the offer set is the line $\boldsymbol{x} = (-K, l > l_0)$, where $-K$ is the fixed amount of capital spent on paying wages in exchange for the amount of labour l greater than a minimum value l_0. According to Eq. (3.40), the model predicts the exponential probability distributions for the wages w received by the workers and for the labour l employed by the firms. In the real economy, the probability distribution of wages can be compared with income distribution, and the distribution of labour employed by firms can be related to the distribution of firm sizes (number of employees). However, Foley (1996) simplifies with an artificial assumption that each firm spends the same amount of capital K on labour, which is unrealistic since there is a large variation in the amount of the firms' capitals, which includes the capital spent on labour. One has to look at the probability distribution of wages, unconditional on the distribution of labour. Keeping a fixed value for $x^{(1)}$, let us take a summation over the values of $x^{(2)}$ in Eq. (3.40), as if integrating out the degree of freedom $x^{(2)} = l$. One thus obtains the unconditional probability distribution of the remaining degree of freedom $x^{(1)} = w$, still to be an exponential:

$$P(w) = c\, e^{-w/T_w}, \tag{3.41}$$

c being the normalization constant, and $T_w = 1/\pi^{(1)}$ is the *wage temperature* (Yakovenko 2012).

How does one satisfy the constraint (3.39) with respect to wages? The N_f firms supply the total capital for wages $W = KN_f$, which enters as a negative term into the sum (3.39). Since the constraint is a global one, we can consider W as an input parameter of the model. Given that the unemployed workers have zero wage

$w = 0$, the constraint (3.39) can be rewritten as

$$\sum_{w_k > w_0} w_k N_k = W, \tag{3.42}$$

the sum being taken over the total number of employed workers N_e. The average wage per employed worker is $\langle w \rangle = W/N_e$. Using Eq. (3.41) and replacing the sum over k with an integration over w in Eq. (3.42), one can easily relate $\langle w \rangle$ and T_w

$$\langle w \rangle = \frac{W}{N_e} = \frac{\int_{w_0}^{\infty} w\, P(w)\, dw}{\int_{w_0}^{\infty} P(w)\, dw}, \qquad T_w = \langle w \rangle - w_0. \tag{3.43}$$

Thus, the wage temperature T_w is the difference in the average wage per employed worker and the minimal wage (Yakovenko 2012).

The model also contains unemployed workers, whose number N_u depends on the measure of statistical weight assigned to the state with $w = 0$. Because this measure is an input parameter of the model, one might as well take N_u as an input parameter. The exponential distribution of wages (3.41) actually matches with the real empirical data on income distribution for the majority of the population.

It should be mentioned in a similar context that, contrary to the conventional picture that the free market for labour, which determines the pay packages, cares only about efficiency and not fairness, Venkatasubramanian (2010) proposed an alternative picture, which shows that an ideal free market environment may also promote fairness. Venkatasubramanian (2010) suggests that this arises as an 'emergent' property resulting from the 'self-organizing' market dynamics (Bak 1996). Although an individual employee may care only about his or her own salary, the collective actions of all the employees, combined with the profit-maximizing actions of each and every company, in a free market environment under budgetary constraints, may lead towards a fair allocation of wages, as professed in Adam Smith's 'invisible hand of self-organization' theory (Smith 1776). He shows that entropy may be considered as an appropriate measure of fairness in a free market environment, which is maximized at equilibrium to yield the log-normal distribution of salaries in an organization under ideal conditions. Thus, the author makes distinction between 'inequality' and 'fairness', and proposes that a fair distribution (that maximizes entropy of the distribution) would have some inequality.

Econophysicists have also considered that maximization of entropy gives rise to the 'natural' inequality in income and wealth distributions, as we shall see in the context of kinetic exchange models, in the following chapters.

4

Market exchanges and scattering process

There has been a considerable development in a special class of asset exchange models, in the last decade or so, which have been formulated in analogy with the kinetic theory of gases. While the essential features are very similar to the gas models, certain aspects modelling economic phenomena give rise to strikingly different results. We will devote this chapter solely to the description of these models and their numerical results.

4.1 Gas-like models

Mandelbrot (1960) wrote, 'There is a great temptation to consider the exchanges of money which occur in economic interaction as analogous to the exchanges of energy which occur in physical shocks between molecules. In the loosest possible terms, both kinds of interactions *should* lead to *similar* states of equilibrium. That is, one *should* be able to explain the law of income distribution by a model similar to that used in statistical thermodynamics: many authors have done so explicitly, and all the others of whom we know have done so implicitly'.

In analogy to two-particle collisions with a resulting change in their individual kinetic energies (or momenta), income exchange models may be based on two-agent interactions. Here two randomly selected agents exchange money by some predefined mechanism. Assuming the exchange process does not depend on previous exchanges, the dynamics follow a Markovian process, which can be represented in general as:

$$\begin{pmatrix} m_i(t+1) \\ m_j(t+1) \end{pmatrix} = \mathcal{M} \begin{pmatrix} m_i(t) \\ m_j(t) \end{pmatrix}, \tag{4.1}$$

where $m_i(t)$ is the income of agent i at time t and the collision matrix $\mathcal{M}$ defines the exchange mechanism (shown in Fig. 4.1).

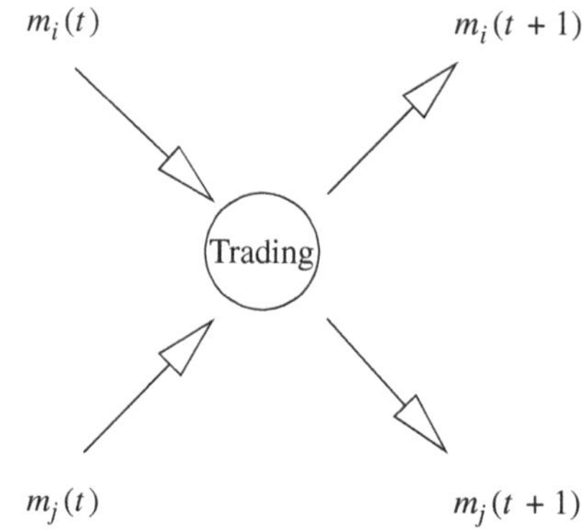

Figure 4.1 The trading process. Agents i and j redistribute their money in the market: $m_i(t)$ and $m_j(t)$, their respective money before trading, changes over to $m_i(t+1)$ and $m_j(t+1)$ after trading.

This class of models considers a 'closed' economic system, in the sense that the total money M and total number of agents N both remain fixed. This corresponds to a situation in which no production or migration occurs and the only economic activity is confined to trading, or the economic growth is very slow compared with the trading, so that the above assumptions roughly hold true. Thus, each agent i, which could be an individual or a corporation, possesses money $m_i(t)$ at time t. In any trading, a pair of traders i and j exchange their money (Chakrabarti and Marjit 1995; Ispolatov *et al.* 1998; Chakraborti and Chakrabarti 2000; Drăgulescu and Yakovenko 2000), such that their total money is (locally) conserved and none ends up with negative money ($m_i(t) \geq 0$, i.e. debt is not allowed):

$$m_i(t+1) = m_i(t) + \Delta m; \; m_j(t+1) = m_j(t) - \Delta m, \tag{4.2}$$

following local conservation:

$$m_i(t) + m_j(t) = m_i(t+1) + m_j(t+1); \tag{4.3}$$

time (t) changes by one unit after each trading. The simplest model considers a random fraction of total money to be shared (Drăgulescu and Yakovenko 2000):

$$\Delta m = \epsilon_{ij}[m_i(t) + m_j(t)] - m_i(t), \tag{4.4}$$

where ϵ_{ij} is a random fraction ($0 \leq \epsilon_{ij} \leq 1$) changing with time or trading. The steady-state ($t \to \infty$) distribution of money in this model (DY model hereafter) is Gibbs:

$$P(m) = (1/T)\exp(-m/T); \quad T = M/N. \tag{4.5}$$

Hence, no matter how uniform or justified the initial distribution is, the eventual steady state corresponds to the Gibbs distribution where most of the people have got very little money. This follows from the conservation of money and additivity of entropy:

$$P(m_1)P(m_2) = P(m_1 + m_2). \tag{4.6}$$

This steady-state result is robust, and actually turns out to be quite realistic. In fact, several variations of the trading, and of the 'lattice' (on which the agents can be put and each agent trades with its 'lattice neighbours' only), whether a completely connected regular graph, small-world or fractal like (Moss de Oliveira *et al.* 1999), leaves the distribution unchanged. However, other variations such as random sharing of an amount $2m_2$ only (not of $m_1 + m_2$) when $m_1 > m_2$ (trading at the level of the lower economic class in a chosen pair) lead to a drastic situation: all the money in the market drifts to one agent, leaving the rest truly pauper (Chakraborti 2002; Hayes 2002).

4.1.1 Model with uniform savings

In any trading, savings come naturally (Samuelson 1998). A saving propensity factor λ was introduced in the random exchange model (Chakraborti and Chakrabarti 2000), in which each trader at time t saves a fraction λ of its money $m_i(t)$ and trades randomly with the rest:

$$m_i(t+1) = \lambda m_i(t) + \epsilon_{ij}\left[(1-\lambda)(m_i(t) + m_j(t))\right], \tag{4.7}$$

$$m_j(t+1) = \lambda m_j(t) + (1-\epsilon_{ij})\left[(1-\lambda)(m_i(t) + m_j(t))\right], \tag{4.8}$$

where

$$\Delta m = (1-\lambda)[\epsilon_{ij}\{m_i(t) + m_j(t)\} - m_i(t)], \tag{4.9}$$

where ϵ_{ij} is a random fraction. By definition, λ is a proper fraction, i.e. $0 \le \lambda \le 1$. This randomness reflects the stochastic nature of the trading. See Drăgulescu and Yakovenko (2000) for the model without savings ($\lambda = 0$).

In this model (referred to hereafter as the CC model), the market (non-interacting at $\lambda = 0$ and 1) becomes 'interacting' for any other non-vanishing λ: for fixed λ (same for all agents), the steady-state distribution $P(m)$ of money is exponentially decaying on both sides with the most-probable money per agent shifting away from $m = 0$ (for $\lambda = 0$) to M/N as $\lambda \to 1$ (Fig. 4.2). This self-organizing feature of the market, induced by sheer self-interest of saving by each agent without any global perspective, is quite significant as the fraction of paupers decreases with saving fraction λ and most people end up with some finite fraction of the average money in the market: for $\lambda \to 1$, the economy is ideally 'socialist', and this is achieved just with people's self-interest of saving. Interestingly, self-organization also occurs in

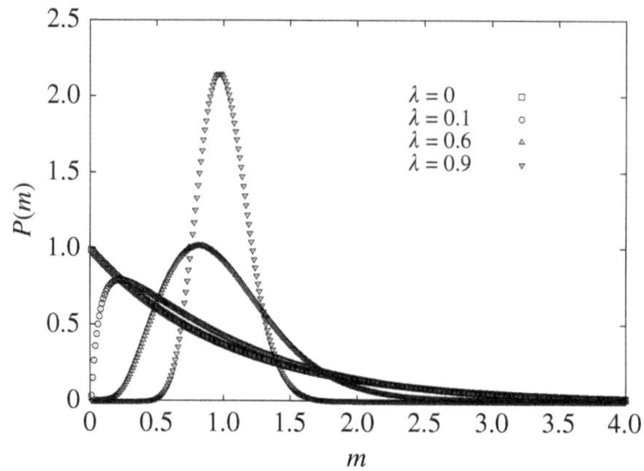

Figure 4.2 Steady-state money distribution $P(m)$ for the model with uniform savings. The data shown are for different values of λ: 0, 0.1, 0.6, 0.9 for a system size $N = 100$. All data sets shown are for average money per agent $M/N = 1$. Reproduced from Chatterjee and Chakrabarti (2007b).

such market models when there is restriction in the commodity market (Chakraborti *et al.* 2001), which we will discuss in the next subsection. Although this fixed saving propensity does not yet give the Pareto-like power law distribution, the Markovian nature of the scattering or trading processes (Eq. (4.6)) is effectively lost. Indirectly through λ, the agents get to know (start interacting with) each other and the system cooperatively self-organizes towards a most-probable distribution ($m_p \neq 0$) (Fig. 4.2).

There have been a few attempts to analytically formulate this problem (Das and Yarlagadda 2003) but no analytic expression has been arrived at. The details of the approach are given in Section 5.1. It has also been claimed through heuristic arguments (based on numerical results) that the distribution is a close approximate form of the gamma distribution (Patriarca *et al.* 2004):

$$P(m) = \frac{n^n}{\Gamma(n)} m^{n-1} \exp(-nm), \tag{4.10}$$

where $\Gamma(n)$ is the gamma function whose argument n is related to the savings factor λ as:

$$n = 1 + \frac{3\lambda}{1-\lambda}. \tag{4.11}$$

The details of the formulation are given in Section 5.1.1. This result has also been supported by numerical results in Bhattacharya *et al.* (2005). However, a later study (Repetowicz *et al.* 2005) analysed the moments, and found that moments up to the third order agree with those obtained from the form of Eq. (4.11), and discrepancies start from fourth order onwards. Hence, the actual form of the distribution for this model still remains to be found out.

It seems that a very similar model was proposed by Angle (1986, 2006) several years back in sociology journals. Angle's 'one parameter inequality process' model (OPIP) is described by the equations:

$$m_i(t+1) = m_i(t) + \mathcal{D}_t w m_j(t) - (1-\mathcal{D}_t) w m_i(t),$$
$$m_j(t+1) = m_i(t) + (1-\mathcal{D}_t) w m_i(t) - \mathcal{D}_t w m_j(t), \tag{4.12}$$

where w is a fixed fraction and $\mathcal{D}_t$ takes value 0 or 1 randomly. The numerical simulation results of OPIP fit well to gamma distributions.

4.1.2 Model with distributed savings

In a real society or economy, the interest of saving varies from person to person, which implies that λ is a very inhomogeneous parameter. To reproduce this situation, we move a step closer to the real situation where saving factor λ is widely distributed within the population (Chatterjee *et al.* 2003, 2004; Chakrabarti and Chatterjee 2004). The evolution of money in such a trading can be written as:

$$m_i(t+1) = \lambda_i m_i(t) + \epsilon_{ij}\left[(1-\lambda_i)m_i(t) + (1-\lambda_j)m_j(t)\right], \tag{4.13}$$
$$m_j(t+1) = \lambda_j m_j(t) + (1-\epsilon_{ij})\left[(1-\lambda_i)m_i(t) + (1-\lambda_j)m_j(t)\right]. \tag{4.14}$$

The trading rules are similar to before, except that

$$\Delta m = \epsilon_{ij}(1-\lambda_j)m_j(t) - (1-\lambda_i)(1-\epsilon_{ij})m_i(t) \tag{4.15}$$

here, where λ_i and λ_j are the saving propensities of agents i and j. In this model (referred to hereafter as the CCM model), the agents have fixed (over time) saving propensities, distributed independently, randomly and uniformly (white) within an interval 0 to 1: agent i saves a random fraction λ_i $(0 \le \lambda_i < 1)$ and this λ_i value is quenched for each agent, i.e. λ_i are independent of trading or t. Starting with an arbitrary initial (uniform or random) distribution of money among the agents, the market evolves with the trading. At each time, two agents are randomly selected and the money exchange among them occurs, following the above-mentioned scheme. We check for the steady state, by looking at the stability of the money distribution in successive Monte Carlo steps t (we define one Monte Carlo time step as N pairwise exchanges). Eventually, after a typical relaxation time the money distribution becomes stationary. This relaxation time is dependent on system size N and the distribution of λ (e.g. $\sim 10^6$ for $N = 1000$ and uniformly distributed λ). After this, we average the money distribution over $\sim 10^3$ time steps. Finally we take the configuration average over $\sim 10^5$ realizations of the λ distribution to

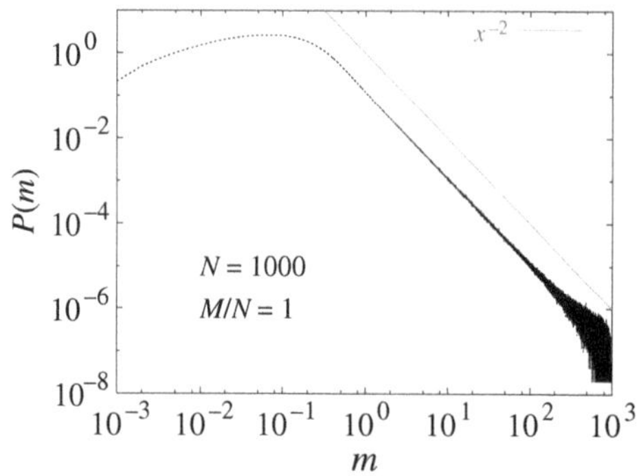

Figure 4.3 Steady-state money distribution $P(m)$ for the distributed λ model with $0 \leq \lambda < 1$ for a system of $N = 1000$ agents. The x^{-2} is a guide to the observed power law, with $1 + \nu = 2$. Here, the average money per agent $M/N = 1$. Reproduced from Chatterjee and Chakrabarti (2007b).

get the money distribution $P(m)$. It is found to follow a strict power law decay. This decay fits to Pareto law (2.2) with $\nu = 1.01 \pm 0.02$ (Fig. 4.3). Note that, for finite size N of the market, the distribution has a narrow initial growth up to a most-probable value m_p after which it falls off with a power law tail for several decades. This Pareto law (with $\nu \simeq 1$) covers almost the entire range in money m of the distribution $P(m)$ in the limit $N \to \infty$. This power law is extremely robust: apart from the uniform λ distribution used in these simulations in Fig. 4.3, this also holds for a distribution

$$\rho(\lambda) \sim |\lambda_0 - \lambda|^{\alpha}, \quad \lambda_0 \neq 1, \quad 0 < \lambda < 1, \tag{4.16}$$

of quenched λ values among the agents. The Pareto law with $\nu = 1$ is universal for all α. The data in Fig. 4.3 correspond to $\lambda_0 = 0$, $\alpha = 0$. For negative α values, however, we get an initial (small m) Gibbs-like decay in $P(m)$ (Fig. 4.4).

In the case of uniformly distributed saving propensity λ ($\rho(\lambda) = 1, 0 \leq \lambda < 1$), the individual money distribution $P(m_k|\lambda_k)$ for an agent with any particular λ_k value, although it differs considerably, remains non-monotonic (Fig. 4.5), similar to that for uniform λ market with $m_p(\lambda)$ shifting with λ (Fig. 4.2). A few subtle points may be noted though: while for uniform λ the $m_p(\lambda)$ were all less than of the order of unity (average money per agent is fixed to $M/N = 1$; Fig. 4.2), for distributed λ case $m_p(\lambda)$ can be considerably larger and can approach the order of N for large λ (Fig. 4.5). There is also a marked qualitative difference in fluctuations (Fig. 4.6): while, for fixed λ, the fluctuations in time (around the most-probable value) in the individuals' money $m_i(t)$ gradually decrease with increasing λ, for quenched distribution of λ, the trend gets reversed (Fig. 4.6).

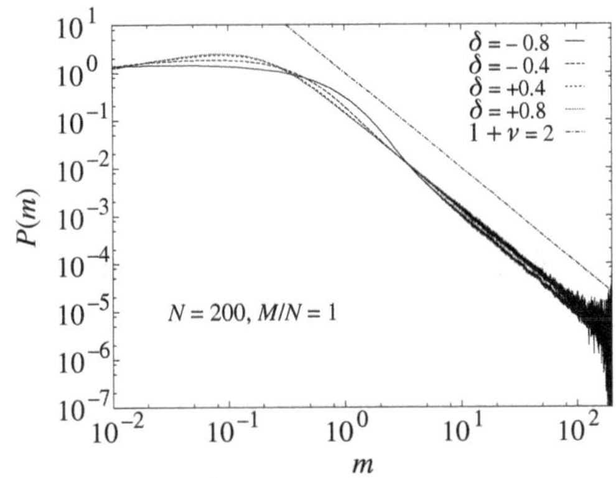

Figure 4.4 Steady-state money distribution $P(m)$ in the model for $N = 200$ agents with λ distributed as $\rho(\lambda) \propto \lambda^{\alpha}$ with different values of α. For all cases, the average money per agent $M/N = 1$. Reproduced from Chatterjee and Chakrabarti (2007b).

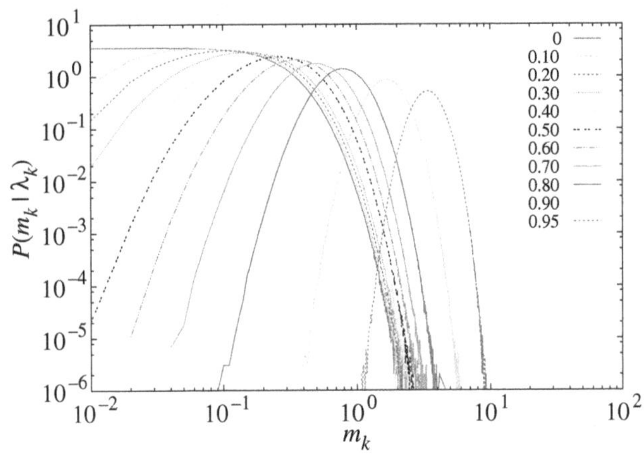

Figure 4.5 Steady-state money distribution $P(m_k|\lambda_k)$ for some 'tagged' agents with typical values of savings λ_k (= 0, 0.10, 0.20, 0.30, 0.40, 0.50, 0.60, 0.70, 0.80, 0.90, 0.95) in the distributed λ model. The data are collected from the ensembles with $N = 256$ agents. For all cases, the average money per agent $M/N = 1$. Reproduced from Chatterjee and Chakrabarti (2007b).

We investigated the range of distribution of the saving propensities in a certain interval $a < \lambda_i < b$, where $0 < a < b < 1$. For uniform distribution within the range, we observed the appearance of the same power law in the distribution but for a narrower region. As may be seen from Fig. 4.7, as $a \to b$, the power law behaviour is seen for values a or b approaching more and more towards unity: for the same width of the interval $|b - a|$, one obtains power law (with the same value of ν) when $b \to 1$. This indicates that, for fixed λ, $\lambda = 0$ corresponds to the

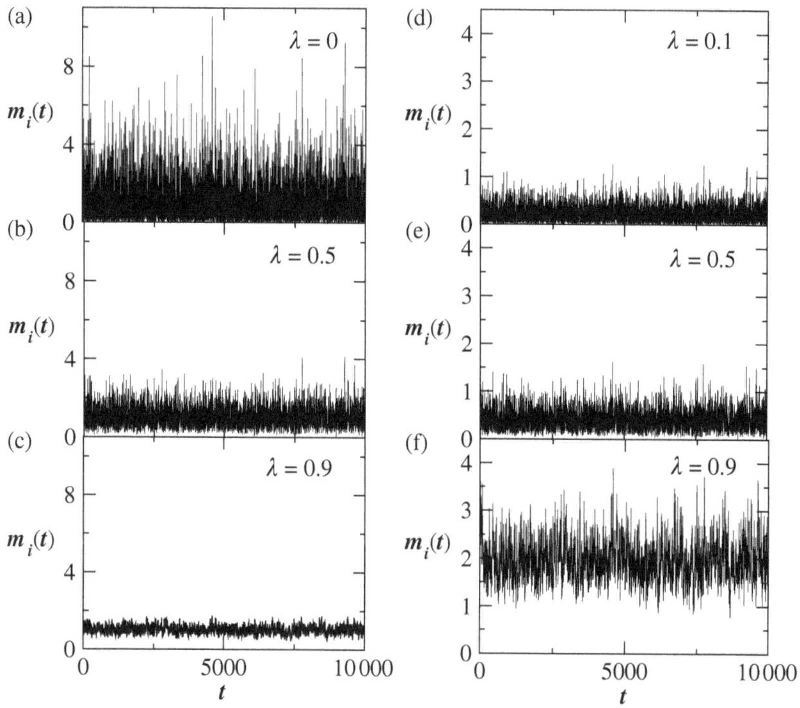

Figure 4.6 Time variation of the money of the ith trader: for uniform λ model, (a–c); and for agents with specific values of λ in the distributed λ model, (d–f). For all cases, the average money per agent $M/N = 1$. Reproduced from Chatterjee and Chakrabarti (2007b).

Gibbs distribution, and one observes a power law in $P(m)$ when λ has got a non-zero width of its distribution extending up to $\lambda = 1$. It must be emphasized at this point that we are talking about the limit $\lambda \to 1$, since any agent having $\lambda = 1$ will result in condensation of money with that particular agent. The role of the agents with high saving propensity ($\lambda \to 1$) is crucial: the power law behaviour is truly valid up to the asymptotic limit if 1 is included. Indeed, had we assumed $\lambda_0 = 1$ in Eq. (4.16), the Pareto exponent ν immediately switches over to $\nu = 1 + \alpha$. Of course, $\lambda_0 \neq 1$ in Eq. (4.16) leads to the universality of the Pareto distribution with $\nu = 1$ (irrespective of λ_0 and α). Obviously, $P(m) \sim \int_0^1 P(m_k|\lambda_k)\rho(\lambda_k)d\lambda_k \sim m^{-2}$ for $\rho(\lambda)$ given by Eq. (4.16) and $P(m) \sim m^{-(2+\alpha)}$ if $\lambda_0 = 1$ in Eq. (4.16) (for large m values).

These theoretical income distributions $P(m)$ compare very well with the empirical distributions of various countries: data suggest a Gibbs-like distribution in the low-income range (more than 90% of the population) and a Pareto-like distribution in the high-income range (Levy and Solomon 1997; Drăgulescu and Yakovenko

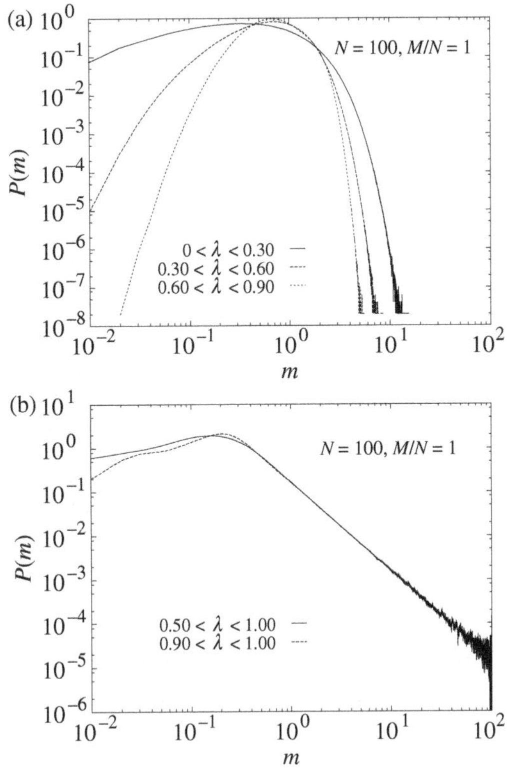

Figure 4.7 Steady-state money distribution in cases when the saving propensity λ is distributed uniformly within a range of values: (a) width of λ distribution is 0.3, money distribution shows a small power law region only for $0.6 < \lambda < 0.9$; (b) λ distribution extends up to 1, but is of different widths: $0.5 < \lambda < 1.0$ and $0.9 < \lambda < 1.0$; money distribution shows power law decay. The power law exponent is $\nu \simeq 1$ in all cases. All data shown here are for $N = 100$, $M/N = 1$. Reproduced from Chatterjee and Chakrabarti (2007b).

2001b; Aoyama *et al.* 2003) (less than 10% of the population) of various countries. In fact, we compared one model simulation of the market with saving propensity of the agents distributed following Eq. (4.16), with $\lambda_0 = 0$ and $\alpha = -0.7$ (Chatterjee *et al.* 2004). The qualitative resemblance of the model income distribution with the real data for Japan and the USA in recent years is quite intriguing. In fact, for negative α values in Eq. (4.16), the density of traders with low saving propensity is higher and, since the $\lambda = 0$ ensemble yields Gibbs-like income distribution (4.5), we see an initial Gibbs-like distribution which crosses over to Pareto distribution (2.2) with $\nu = 1.0$ for large m values. The position of the crossover point depends on the value of α. It is important to note that any distribution of λ near $\lambda = 1$, of finite width, eventually gives Pareto law for large m limit. The same kind of

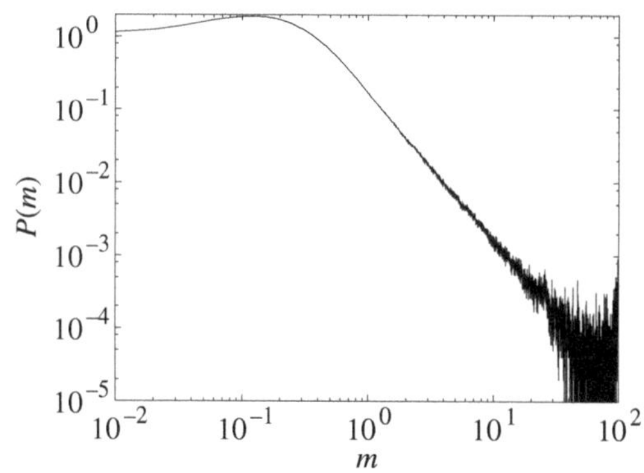

Figure 4.8 Distribution $P(m)$ of money m in case of annealed savings λ varying randomly in $[\mu, 1)$. Here, $\zeta(\mu)$ has a uniform distribution. The distribution produces a power law tail with Pareto exponent $\nu = 1$. The simulation has been done for a system of $N = 100$ agents, with average money per agent $M/N = 1$. $P(m)$ is the steady-state distribution after 4×10^4 Monte Carlo steps, and the data are averaged over an ensemble of 10^5. Taken from Chatterjee and Chakrabarti (2007a).

crossover behaviour (from Gibbs to Pareto) can also be reproduced in a model market of mixed agents where $\lambda = 0$ for a finite fraction p of population and λ is distributed uniformly over a finite range near $\lambda = 1$ for the rest $1 - p$ fraction of the population.

4.1.3 Model with 'annealed' savings

In a real trading process, the concept of a 'saving factor' cannot be attributed to a quantity that is invariant with time. A saving factor is most likely to change with time or trading. The case of annealed savings in the distributed savings model, in which the savings factor λ_i changes with time in the interval [0, 1), does not produce a power law in $P(m)$ (Chatterjee *et al.* 2004). But, there are some special cases of annealed saving which can produce a power law distribution of $P(m)$.

If one allows the saving factor λ_i to vary with time in [0, 1), the money distribution $P(m)$ does not produce a power law tail (Chatterjee *et al.* 2004). Instead, one can conceive a slightly different model of an annealed saving case. We assigned a parameter μ_i $(0 < \mu_i < 1)$ to each agent i, such that the savings factor λ_i randomly assumes a value in the interval $[\mu_i, 1)$ at each time or trading. The trading rules are of course unaltered and governed by Eqs. (4.13) and (4.14). Now, considering a suitable distribution $\zeta(\mu)$ of μ over the agents, one can produce money distributions with a power law tail. The only condition that needs to be satisfied is that $\zeta(\mu)$ should be non-vanishing as $\mu \to 1$. Figure 4.8 shows the case when $\zeta(\mu) = 1$.

Numerical simulations suggest that the behavior of the wealth distribution is similar to the quenched savings case. In other words, only if $\zeta(\mu) \propto |1-\mu|^{\alpha}$, it is reflected in the Pareto exponent as $\nu = 1+\alpha$ (Chatterjee and Chakrabarti 2007a). μ_i is interpreted as the lower bound of the saving distribution of the i-th agent. Thus, while agents are allowed to randomly save any fraction of their money, the bound ensures that there is always a non-vanishing fraction of the population that assumes high saving fraction.

4.1.4 Further studies

Further studies have provided deeper insight into the structure of the models and their solutions.

4.1.4.1 Correlation between savings and average money

Patriarca *et al.* (2005) studied the correlation between the saving factor λ and the average money held by an agent whose savings factor is λ. This numerical study revealed that the product of this average money and the unsaved fraction remains constant, or, in other words, the quantity $\langle m(\lambda)\rangle(1-\lambda)$ is a constant. This result turns out to be the key to the formulation of a mean-field analysis to the model (Mohanty 2006), which will be discussed in detail in the following chapter.

Another numerical study (Bhattacharya *et al.* 2005) analysed the average money of the agent with the maximum savings factor $\langle m(\lambda_{\max})\rangle$. This study concludes on the time evolution of the money of this agent, and finds a scaling behaviour

$$[\langle m(\lambda_{\max})\rangle/N]\,(1-\lambda_{\max})^{0.725} \sim \mathcal{G}\,[t(1-\lambda_{\max})]\,. \tag{4.17}$$

This implies that the stationary state for the agent with the maximum value of λ is reached after a relaxation time

$$t_{\times} \propto (1-\lambda_{\max})^{-1}. \tag{4.18}$$

The average money $\langle m(\lambda_{\max})\rangle$ of this agent is also found to scale as

$$[\langle m(\lambda_{\max})\rangle/N]N^{-0.15} \sim \mathcal{F}[(1-\lambda_{\max})N^{1.5}]. \tag{4.19}$$

The scaling function $\mathcal{F}[x] \to x^{-\beta}$ as $x \to 0$ with $\beta \approx 0.76$. This means $\langle m(\lambda_{\max})\rangle N^{-1.15} \sim (1-\lambda_{\max})^{-0.76}N^{-1.14}$ or $\langle m(\lambda_{\max})\rangle \sim (1-\lambda_{\max})^{-0.76}N^{-0.01}$. Since for a society of N traders $(1-\lambda_{\max}) \sim 1/N$ this implies

$$\langle m(\lambda_{max})\rangle \sim N^{0.77}. \tag{4.20}$$

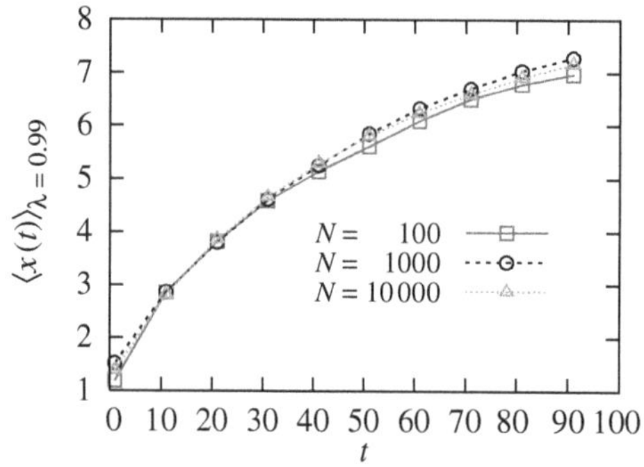

Figure 4.9 Average wealth $\langle x(t)\rangle_{\lambda=0.99}$ versus the rescaled time $t = T/N$ for systems with different numbers of agents $N = 100$, 1000, and 10 000, but the same saving propensity density $\phi(\lambda)$ that uniformly partitions agents into 100 subsets with $\lambda = 0.00,\ 0.01, \ldots, 0.99$. T is the total number of trades. Reproduced from Patriarca *et al.* (2007).

4.1.4.2 Relaxation studies

Relaxation process If a real economic system is characterized by a wealth distribution with a certain shape, it is of great interest to know on which time scale the system relaxes towards this distribution from a given arbitrary initial distribution of wealth, and how the relaxation process depends on the system parameters, in particular on the system size and the distribution of saving propensities. Patriarca *et al.* (2007) made such studies by numerical simulations.

In the simulations, all agents started from the same initial wealth $x_i(t = 0) = x_0 = 1$, without any loss of generality. The value x_0, owing to the conservation of the total wealth $X = \sum_{i=1}^{N} x_i$, also represented the global average value of x at any time t, i.e. $\langle x(t)\rangle \equiv \int x f(x)\,dx = x_0 = X/N$. This set-up was used to model a more general situation in which the initial conditions of the agents were far from equilibrium.

Relaxation to equilibrium as a function of system size Before analysing the dependence of the time scale on the saving propensity distribution, Patriarca *et al.* (2004) considered its dependence on the number of agents N. If time was measured by the number of transactions T, they found that the time scale was proportional to the number of agents N: a system A that is m times larger than a system B ($N_A = mN_B$) relaxes m times slower than B. This is shown in Fig. 4.9, where the average wealth $\langle x(t)\rangle_\lambda$ of the agent subset with $\lambda = 0.99$ is plotted for various systems with different values of N versus the rescaled time $t = T/N$. However, the λ-density $\phi(\lambda)$ is the same for all systems and uniformly partitions each system into 100 subsets with values $\lambda = 0.00,\ 0.01, \ldots, 0.99$.

Time t is defined as the ratio $t = T/N$ between the total number of trades T and the total number of agents N, i.e. what is usually called a Monte Carlo cycle or sweep in molecular simulation (Allen and Tildesley 1989): in a Monte Carlo cycle, each agent performs on average the same number of trades (actually two), in the same fashion as in molecular dynamics – each particle is moved once at every time step. The results do not change if one of the two agents involved in an exchange is selected sequentially, e.g. in the order of its index $i = 1, \ldots, N$, as is common practice in molecular simulations. This ensured that every agent performs at least one trade per cycle and reduces the amount of random numbers to be drawn. They introduced another time unit τ_0, such that during any time interval $(t, t + \tau_0)$ *all* agents perform on average one trade (or the same number of trades). In this way the dynamics and the relaxation process became independent of N. The existence of a natural time scale independent of the system size provided a foundation for using simulations of systems with finite N in order to infer properties of systems with continuous saving propensity distributions and $N \to \infty$.

Relaxation to equilibrium as a function of saving propensity Relaxation in systems with constant λ was studied by Chakraborti and Chakrabarti (2000), where a systematic increase of the relaxation time with λ, and eventually a divergence for $\lambda \to 1$, was found: for $\lambda = 1$, no exchanges can occur, so that the system is frozen.

Patriarca *et al.* (2007) considered systems with uniformly distributed λ. In this case a similar behaviour of the relaxation times was observed, broken down to subsystems with similar values of λ. As discussed in detail in Patriarca *et al.* (2005, 2006) and Bhattacharya *et al.* (2005), the partial wealth distributions of agents with a given value of λ relax towards different states with characteristic shapes $f_\lambda(x)$. The generic function $f_\lambda(x)$ has a maximum and an exponential tail, thus closely recalling the shape of a Γ-distribution. The corresponding average value is given by $\langle x \rangle_\lambda \equiv \int x f_\lambda(x)\, dx = k/(1 - \lambda)$, where k is a suitable constant determined through the condition $\int \langle x \rangle_\lambda \phi(\lambda)\, d\lambda = X/N$; X is the total wealth of the system. Even if the partial distributions decay exponentially with x, the sum of all partial distributions results in a Pareto law at large values of x, i.e. $f(x) = \sum_\lambda f_\lambda(x) \sim 1/x^{1+\alpha}$. Numerical simulations clearly show that agents with different values of λ are associated to different relaxation times τ_λ.

Results are illustrated in Fig. 4.10 for a system of $N = 10^4$ agents uniformly partitioned into 100 subsets with $\lambda = 0.01,\ 0.02, \ldots, 0.99$: mean wealths of subsets corresponding to a value of λ closer to 1 relax slower towards their asymptotic average wealth $\langle x \rangle_\lambda \propto 1/(1 - \lambda)$.

The average wealth x_0 allowed to introduce a threshold that partitions the system into *poor agents*, with an asymptotic average wealth $\langle x(t \to \infty) \rangle_\lambda < x_0$, and

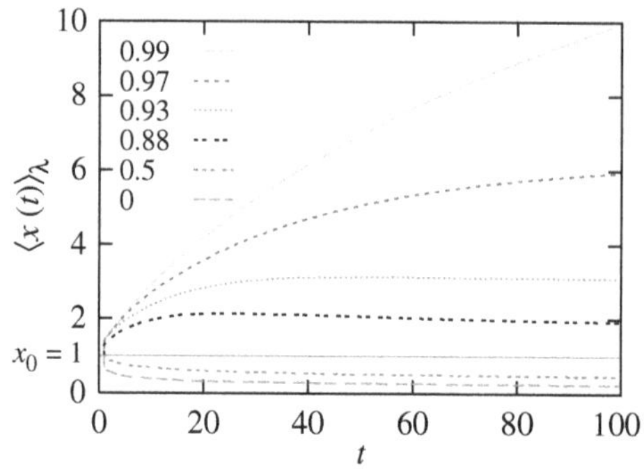

Figure 4.10 Mean wealth $\langle x(t) \rangle_\lambda$ versus time for various λ. The 10^4 agents are uniformly partitioned into 100 subsets with $\lambda = 0.01,\ 0.02, \ldots, 0.99$. Higher λ correspond to longer relaxation times. The continuous line $\langle x \rangle = x_0$ partitions agents into *poor* ($x < x_0$) and *rich* ($x > x_0$) ones. Reproduced from Patriarca *et al.* (2007).

rich agents with $\langle x(t \to \infty) \rangle_\lambda > x_0$. The poor–rich threshold $\langle x \rangle = x_0 = 1$ is represented as a continuous line in Fig. 4.10 and corresponds to $\lambda \approx 0.75$ for the particular example.

The differences in the relaxation process can be related to the different relative wealth exchange rates that, by direct inspection of Eqs. (4.14) and (4.15), appear to be proportional to $1 - \lambda$. Thus, in general, higher saving propensities are expected to be associated with slower relaxation processes. A more detailed analysis was carried out as shown in Fig. 4.11: after the rescaling of time and wealth by the factor $(1 - \lambda)$, mean wealths corresponding to agents with different values of λ (Fig. 4.11a) appeared to relax approximately *on the same time scale* and towards the same asymptotic value (Fig. 4.11b). In fact, the factor $(1 - \lambda)$ was proportional to the wealth exchange rates and, at the same time, through the condition of stationarity, determined the equilibrium average wealth values $\langle x \rangle_\lambda = k/(1 - \lambda)$ (Patriarca *et al.* 2005). As in some earlier cases, agents started from the same initial condition $x_i(t = 0) = x_0 = 1$. In this case, in order to study in greater detail the high saving propensity parameter region, which corresponds to the high relaxation time region, the system of $N = 10^4$ agents was uniformly partitioned into 200 subsets with saving propensities $\lambda = 0.5000,\ 0.5025, \ldots, 0.9975$. It was not strictly a uniform distribution of λ on $[0, 1)$, since $\phi(\lambda) = 0$ for $\lambda < 0.5$; however, it did not matter because what counted was the high saving propensity parameter interval.

Relaxation time distribution The model with distributed saving propensities was completely specified by the trading rules of Eqs. (4.14) and the set of saving propensities $\{\lambda_i\}$ of the N agents. In the case of a continuously distributed λ, a

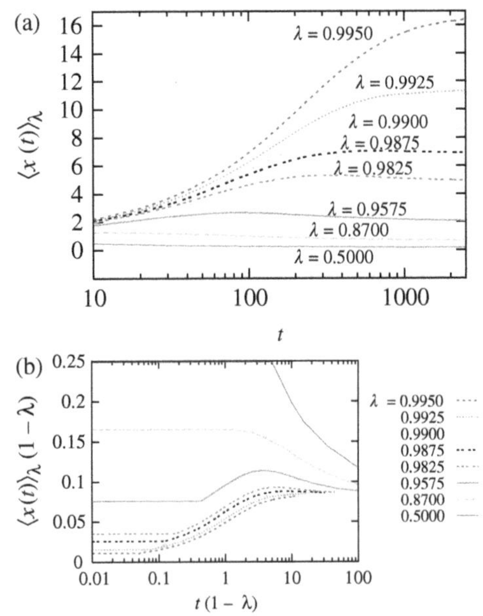

Figure 4.11 (a) Average wealth $\langle x(t)\rangle_\lambda$ versus (log of) time for the λ's listed. The 10^4 agents are partitioned into 200 subsets with $\lambda = 0.5000,\ 0.5025, \ldots, 0.9975$. (b) Same as in (a) after rescaling wealth and time by the factor $(1-\lambda)$, inversely proportional to the mean wealth and proportional to the average wealth exchange rate. Reproduced from Patriarca *et al.* (2007).

continuous saving propensity density $\phi(\lambda)$ was used in place of the discrete λ-set, normalized so that $\int_0^1 \phi(\lambda)\, d\lambda = 1$.

Patriarca *et al.* 2007 suggested a method to obtain the wealth as well as the relaxation distribution directly from the saving propensity density $\phi(\lambda)$. It follows from probability conservation that $\tilde{f}(\bar{x})d\bar{x} = \phi(\lambda)d\lambda$, where $\bar{x}$ is a short notation for $\langle x\rangle_\lambda$ and $\tilde{f}(\bar{x})$ is the density of the average wealth values. In the case of uniformly distributed saving propensities, one obtains

$$\tilde{f}(\bar{x}) = \phi(\lambda)\frac{d\lambda(\bar{x})}{d\bar{x}} = \phi\left(1 - \frac{k}{\bar{x}}\right)\frac{k}{\bar{x}^2}\,. \tag{4.21}$$

This shows that a uniform saving propensity distribution leads to a power law $\tilde{f}(\bar{x}) \sim 1/\bar{x}^2$ in the (average) wealth distribution. In general a λ-density going to zero for $\lambda \to 1$ as $\phi(\lambda) \propto (1-\lambda)^{\alpha-1}$ (with $\alpha \geq 1$) leads to the Pareto law $\tilde{f}(\bar{x}) \sim 1/\bar{x}^{1+\alpha}$ with Pareto exponent $\alpha \geq 1$ as found in real distributions.

In a very similar way it was possible to obtain the associated distribution of relaxation times $\psi(\tau)$ for the global relaxation process through the relation between the relaxation time τ_λ and the agent saving propensity: given that the time scale follows a relation $\tau_\lambda \propto 1/(1-\lambda)$, then

$$\psi(\tau) = \phi(\lambda)\frac{d\lambda(\tau)}{d\tau} \propto \phi\left(1 - \frac{\tau'}{\tau}\right)\frac{\tau'}{\tau^2}, \tag{4.22}$$

where τ' is a proportionality factor. Comparison with Eq. (4.21) shows that $\psi(\tau)$ and $\tilde{f}(\bar{x})$ are characterized by power law tails in τ and $\bar{x}$, respectively, *with the same Pareto exponent.*

It is to be noted, as discussed in Patriarca *et al.* (2005), that in the parameter region $\lambda \to 1$, from which the main contributions to the Pareto power law tail come, the widths of the generic equilibrium partial distributions $f_\lambda(x)$ increase more slowly than the difference between the mean values $\langle x \rangle_{\lambda'} - \langle x \rangle_\lambda$ corresponding to two agents with consecutive values of the saving propensity λ' and λ. This implies that at equilibrium and in the tail of the distribution it is possible to resolve the mixture $\sum_\lambda f_\lambda(x)$ into its components $f_\lambda(x)$ and to approximate the current value of wealth $x(t)$ of a certain agent with saving propensity λ (that is actually a stochastic process) with the corresponding average value, $\langle x \rangle_\lambda \approx x$, so that $\tilde{f}(x) \approx f(x)$.

Finally, Patriarca *et al.* (2007) noted that an ensemble with a power law distribution of relaxation times undergoes a slow relaxation process if the exponent of the relaxation time distribution is smaller than 2, so that a Pareto exponent larger than 2, as automatically generated by the model, seems to ensure a normal relaxation.

4.1.5 Dynamics of agents

In the DY and CC models, agents are homogeneous. The DY model is nothing but a special case of the CC model where $\lambda = 0$. In these models, looking at individual agents and the whole system are equivalent. On the contrary, the presence of the distributed saving propensity (quenched disorder) in the CCM model gives it a rich structure and calls for a careful look at the local scale (at the level of individuals) besides computing global quantities.

Chatterjee and Sen (2010) reported extensive numerical simulations with a system of N agents, with uniform distribution $\Lambda(\lambda) = 1$, bounded above by $1 - 1/N$. One observes the dynamics of a tagged agent k, having a saving propensity λ_k, in a pool of N agents distributed according to a quenched $\Lambda(\lambda)$, to try to see how the individual distributions $P(m_k|\lambda_k)$ look (Fig. 4.5). As reported elsewhere (Chatterjee *et al.* 2004; Bhattacharya *et al.* 2005; Patriarca *et al.* 2005; Chatterjee and Chakrabarti 2007b), the agents with smaller values of savings λ_k barely have money of the order of average money in the market. On the other hand, agents with high saving propensity λ_k possess money comparable to the average money in the market, and in fact, for the richest agent, the distribution extends almost up to the total money M.

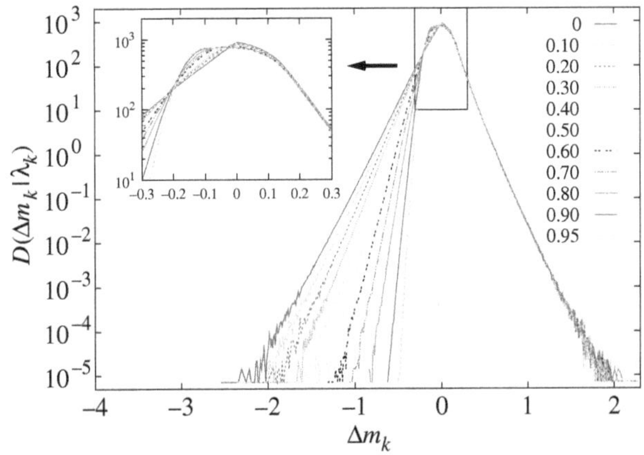

Figure 4.12 Distribution $D(\Delta m_k|\lambda_k)$ of money difference Δm_k for the tagged agent k with a particular value of savings λ_k in the CCM model with uniformly distributed savings ($\delta = 0$). The data are shown for a system of $N = 256$ agents. The inset shows that, for higher λ, the probability of losses becomes larger in the region $-0.2 < \Delta m_k < 0$. Reproduced from Chatterjee and Sen (2010).

4.1.5.1 Distribution of change in wealth

Upon trading with another agent l, the money of the tagged agent k changes by an amount

$$\Delta m_k = m_k(t+1) - m_k(t) = -\left(m_l(t+1) - m_l(t)\right).$$

The distributions $D(\Delta m_k|\lambda_k)$ were computed in the steady state, given that agent k has a saving propensity λ_k (Fig. 4.12). This distribution has asymmetries for both small and large values of saving propensities λ_k.

Total money remains constant in the steady state for any agent. An agent with a relatively higher λ incurs losses which are considerably small in magnitude, immediately suggesting that agents with larger savings must be having more exchanges where losses, however small, occur. The magnified portion of the distribution shows that it is really so (shown in the inset of Fig. 4.12).

4.1.5.2 Walk in the wealth space: definition

To investigate the dynamics at the microscopic level, one can conceive of a walk for the agents in the so-called 'wealth space', in which each agent walks a step forward when she gains and one step backwards if she incurs a loss. The walks are correlated in the sense that when two agents interact, if one takes a step forward, the other has to move backward. On the other hand, two agents can interact irrespective of their positions in the wealth space unlike Brownian particles.

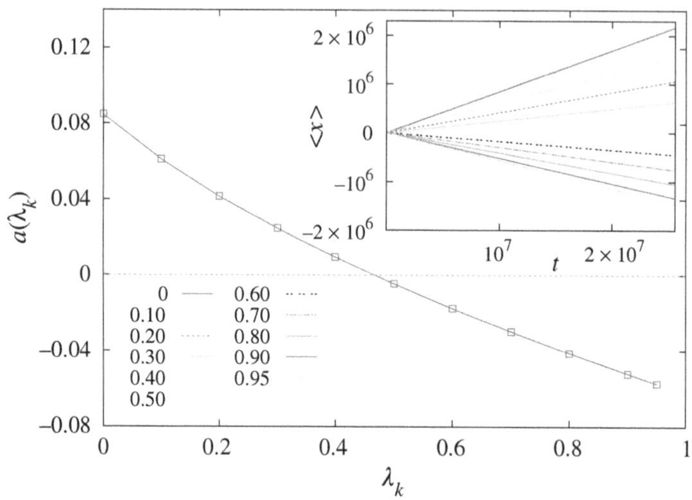

Figure 4.13 Measures for the gain–loss walk: the inset shows $\langle x \rangle$ with time for different values of savings λ_k, showing the drifts. The slopes $a(\lambda_k)$ are also shown. The estimate of λ_k^* is approximately 0.469. The data are shown for a system of $N = 256$ agents. Reproduced from Chatterjee and Sen (2010).

Once the system is in the steady state, one can define $x(t+1) = x(t) + 1$ if the tagged agent gains money, and $x(t+1) = x(t) - 1$ if she loses. In other words, $x(t)$ performs a walk in one dimension. Without loss of generality we start from origin ($x(0) = 0$), and we insist $t = 0$ is well within the steady state. We investigate the properties of this walk by computing the mean displacement $\langle x(t) \rangle$, and the mean square displacement $\langle x^2(t) \rangle - \langle x(t) \rangle^2$.

Actually, one can also consider a walk for a tagged agent in which the increments (i.e. step lengths) are the money gained or lost at each step, but the exponential distribution obtained for such step lengths (Fig. 4.12) indicates that it will be simple diffusion like. This walk in gain–loss space will be referred to as GLS walk hereafter.

For the CC model, for any value of the fixed saving propensity λ, one obtains a conventional random walk in the sense $\langle x(t) \rangle$ is zero and $\langle x^2(t) \rangle - \langle x(t) \rangle^2 \sim t$. However, for the CCM model, results are quite different. It is found that $\langle x(t) \rangle$ has a drift, $\langle x(t) \rangle \sim a(\lambda_k)t$. $a(\lambda_k)$ varies continuously with λ_k, taking positive to negative values as one goes from low to high values of savings λ_k (see inset of Fig. 4.13), respectively. It is obvious that, for some λ_k^*, there is no drift, $a(\lambda_k^*) = 0$. λ_k^* is estimated to be about 0.469 by the interpolation method. On the other hand $\langle x^2 \rangle - \langle x \rangle^2 \sim t^2$ for all λ_k, which is a case of ballistic diffusion (Fig. 4.14). The negative or positive drifts of the walks indicate that the probabilities of gain and loss are not equal for any agent in general. Plotting the fraction of times the tagged

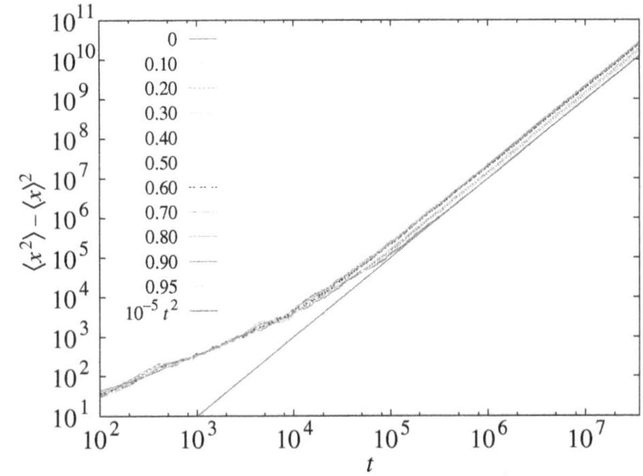

Figure 4.14 Measures for the gain–loss walk: time variation of $\langle x^2\rangle - \langle x\rangle^2$ for different values of savings λ_k, and a guide to t^2. The data are shown for a system of $N = 256$ agents. Reproduced from Chatterjee and Sen (2010).

agent gains/loses in Fig. 4.15a, it is indeed found that an agent with a smaller λ gains with more probability while the opposite happens for agents with larger λ. Indeed, the intersection of the two curves is the point λ_k^* where the probabilities are equal and the corresponding walk should show $\langle x\rangle = 0$. It is difficult, however, to detect numerically λ_k^* exactly, which lies close to 0.47, and check whether an agent with λ_k^* behaves like a conventional random walker or shows ballistic diffusion. Simulations using values of λ even very close to 0.47 always show ballistic behaviour.

In order to explain the above results, one could investigate at a finer level the walk when the tagged agent with λ_k interacts with another agent with saving λ. First, one can calculate the average $\langle\lambda\rangle$ when the tagged agent loses or gains and note that, for a gain, one has to interact with an agent with a higher λ in general, as indicated in Fig. 4.15b. In fact, the average value is very weakly dependent on λ_k and significantly greater/less than 0.5 for a gain/loss. This is contrary to the expectation that gain/loss does not depend on the saving propensities of the interacting agents.

Having obtained evidence that gain/loss depends on the interacting agents' saving propensities, one computes the probability of gain and loss, P_g and P_l, respectively, as a function of λ for the agent with saving λ_k. The data show that indeed an agent gains with higher probability while interacting with an agent with $\lambda > \lambda_k$ and vice versa. In fact, the data for different λ_k collapse when $P_g - P_l$ are plotted against a scaled variable $y = \frac{\lambda-\lambda_k}{1.5+\lambda_k+\lambda}$ as shown in Fig. 4.16 indicating a

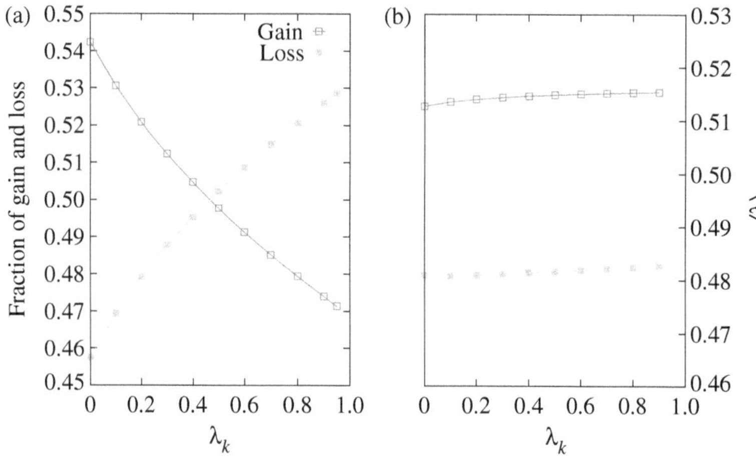

Figure 4.15 (a) Plot of probabilities of gain and loss for different values of savings propensity λ_k of the tagged agent. The data are shown for a system of $N = 256$ agents. (b) The plot of the average value $\langle\lambda\rangle$ when a gain/loss is being incurred shown against λ_k of the tagged agent. Reproduced from Chatterjee and Sen (2010).

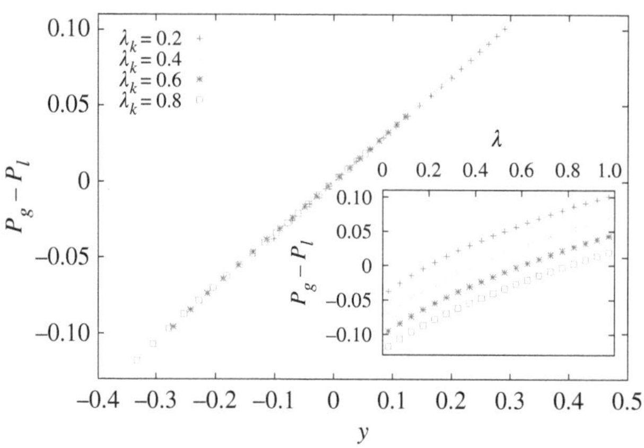

Figure 4.16 Data collapse for $\lambda_k = 0.2, 0.4, 0.6$ and 0.8 is shown for $P_g - P_l$ versus the scaled variable $y = \frac{\lambda - \lambda_k}{1.5+\lambda_k+\lambda}$ for $N = 256$. The inset shows the unscaled data. Reproduced from Chatterjee and Sen (2010).

linear variation with y, i.e.

$$P_g - P_l = \text{const}\frac{\lambda - \lambda_k}{1.5 + \lambda_k + \lambda}. \tag{4.23}$$

It is also observed that there is hardly any finite size effect on the collapse in the sense that the similarly scaled data for $N = 100$ collapse exactly on those for

$N = 256$. An agent with a high value of λ will interact with a higher probability with agents whose saving propensities are lesser, causing a loss of money. Therefore in the wealth space it will have tendency to take more steps in the negative direction. This explains the negative drift for large λ.

It is possible to estimate the value of λ_k^* using Eq. (4.23), utilizing the fact that the integrated value of $P_g - P_l$ over all λ should be zero for $\lambda_k = \lambda_k^*$. This gives

$$1 - (1.5 + 2\lambda_k^*)\left[\log(2.5 + \lambda_k^*) - \log(1.5 + \lambda_k^*)\right] = 0, \tag{4.24}$$

solving which one can get $\lambda_k^* \simeq 0.4658$, which is consistent with the earlier observations.

It may be added here that in principle the probability of gain or of loss while two agents interact can be calculated from the money distribution. In the CCM model, when two agents with money m_1 and m_2 and saving propensities λ_1 and λ_2, respectively, interact, the difference in money before and after the transaction for, say, the second agent is given by $[(1 - \lambda_1)m_1 - (1 - \lambda_2)m_2]/2$. Therefore, for the second agent to lose, m_2 must be greater than $m' = \frac{m_1(1-\lambda_1)}{(1-\lambda_2)}$, and the corresponding probability is given by

$$\int_0^M P(m_1|\lambda_1)dm_1 \int_{m'}^M P(m_2|\lambda_2)dm_2. \tag{4.25}$$

However, the exact form of the money distribution is not known (Basu and Mohanty 2008) for the CCM case. For the CC model, $\lambda = \lambda_1 = \lambda_2$ and, letting $M \to \infty$, the above integral becomes

$$\int_0^\infty P(m_1|\lambda)dm_1 \int_{m_1}^\infty P(m_2|\lambda)dm_2 = \int_0^\infty P(m_1|\lambda)[1 - \tilde{P}(m_1|\lambda)]dm_1, \tag{4.26}$$

where $\tilde{P}(m) = \int_0^m P(m)dm$ is the cumulative distribution of money. Since $P = \frac{\partial \tilde{P}}{\partial m}$, the right-hand side of Eq. (4.26) is equal to $1/2$ independent of the form of $P(m|\lambda)$. Thus, in the CC case, the probability of gain or loss is equal, leading to a simple random walk. In the CCM, however, the results are expected to be dependent on λ_1, λ_2, as Eq. (4.25) indicates.

The above results naïvely suggest that the walk in the GLS is like a biased random walk (BRW) (except perhaps at λ^*) for the CCM model, while it is like a random walk (RW) for the CC model. In fact, in the CCM model, associated with each value of λ, there seems to be a unique value of the parameter p characterizing the corresponding biased random walk, where p is the probability of moving towards a particular direction. This makes it convenient to compare the CCM walk with a BRW, which we discuss in the next subsection by considering some additional features of the walk.

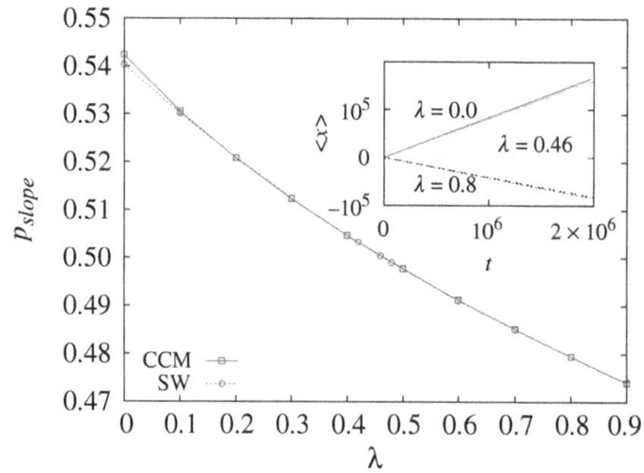

Figure 4.17 Plot of p_{slope} as a function of λ obtained from the slopes of $\langle x \rangle$ versus t plot for the CCM model for $N = 256$ and the simulated walker (SW). Inset shows that the variation of $\langle x \rangle$ against t for $\lambda = 0.0, 0.46$ and 0.8 for the CCM model and the SW are almost indistinguishable. Reproduced from Goswami *et al.* (2011).

4.1.5.3 CCM walk in the GLS: comparison with BRW

In which aspects do the walk in GLS in the CCM model differ from the BRW? In a BRW, a walker moves towards a particular direction with probability $p \neq 1/2$ such that the total distance $\langle x \rangle$ travelled is linear in time t, precisely $\langle x \rangle = (2p - 1)t$ in the preferred direction. To compare the CCM walk with the BRW, Goswami *et al.* (2011) uses the following scheme.

First, one extracts effective values of p for the walk in the GLS using the slopes of the $\langle x \rangle$ versus t plots assuming it is a BRW. Next, from the distribution of distances travelled without a change in direction in the CCM walk, effective p values are extracted assuming it is a BRW. Then, it is possible to compare the direction reversal probability of the CCM walk with that of the BRW. If these effective values of p and direction reversal probability of the CCM walk and the BRW turn out to be identical, one can conclude that the walks in the GLS for the CCM model are ordinary biased random walks.

p using slopes of $\langle x \rangle$ versus t curves As already seen, for the CCM model, $\langle x \rangle$, the distance travelled in time t varies linearly with t. Thus, $\langle x \rangle = s_0 t$ and an effective $p \equiv p_{slope}$ can be calculated using the relation $p_{slope} = \frac{s_0+1}{2}$. (By our convention, if $p_{slope} > \frac{1}{2}$, the walker has a bias towards the right (gain).) The results obtained for this (Fig. 4.17) indicated that p_{slope} approaches $\frac{1}{2}$ as $\lambda \to \lambda^*$.

Distribution of distances travelled without change in direction It is possible to study the distribution of the walk lengths X through which the walker travels

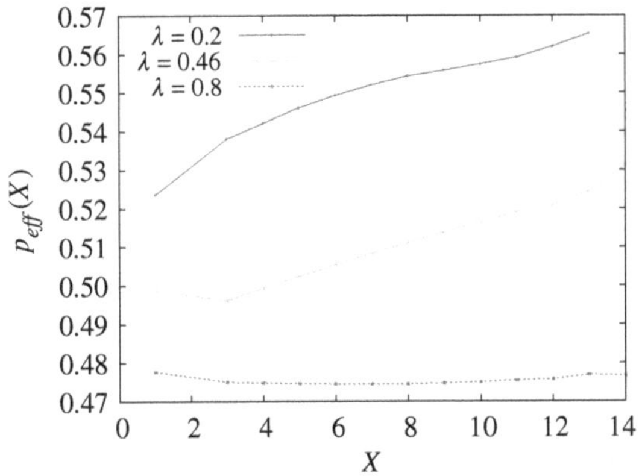

Figure 4.18 Variation of $p_{eff}(X)$ against X for $\lambda = 0.2, 0.46, 0.8$ for the CCM model. p_{eff} values are not independent of X in general. Reproduced from Goswami *et al.* (2011).

without any change in direction. For the BRW, this is easy to calculate: in our convention let the probability to move towards the right be p, then the probability $W_s(X)$ that a walker goes through a length X at a stretch along the right direction is proportional to $p^X(1-p)^2$. The corresponding probability along the left is written as $W_s(-X) \propto p^2(1-p)^X$. Therefore, in a BRW,

$$\frac{W_s(X)}{W_s(-X)} = \left(\frac{p}{1-p}\right)^{X-2}. \tag{4.27}$$

For the walk in the GLS, $W_s(X)/W_s(-X)$ was calculated numerically for any value of λ, and a value of $p_{eff}(X, \lambda)$ for different values of X is obtained using Eq. 4.27. If the CCM walkers are really simple biased random walkers, one would get a $p_{eff}(X, \lambda)$ *independent* of X for a given λ and close to the value p_{slope} obtained using the slope method. Figure 4.18 shows the plot of $p_{eff}(X)$. It should be noted that, in this method, $p_{eff}(X = 2)$ cannot be obtained as the right-hand side of Eq. 4.27 becomes unity, i.e. p independent. We notice immediately that the effective p values are in no way independent of X (except perhaps when λ is close to unity). This strongly indicates that the walks are not simple BRWs. We will get back to this issue in Section 4.1.5 again.

Probability of direction reversal Another quantity closely related to the measure discussed in the previous subsection is the probability of direction reversals made by the walker, which is defined as $f = n_d/n$, where n_d is the number of times the walker changes direction and n is the total number of steps (duration of the walk).

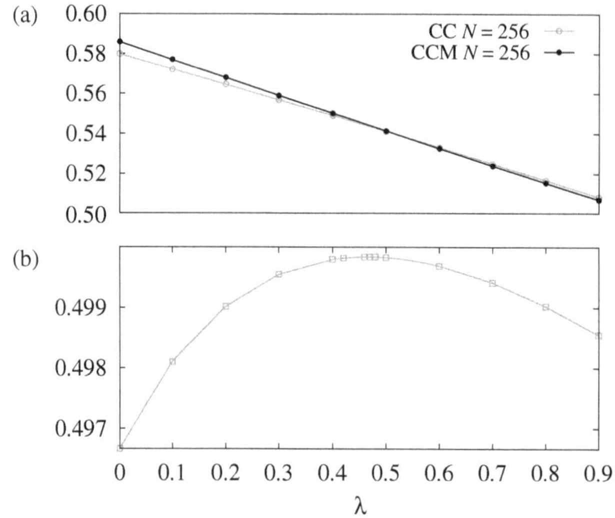

Figure 4.19 (a) Plot of direction reversal probability f against λ for the CCM and CC models. (b) Same for the simulated walk. Reproduced from Goswami *et al.* (2011).

f can be identified as $1/\langle X \rangle$, where

$$\langle X \rangle = \sum_X \left[X W_s(X) + X W_s(-X) \right] \tag{4.28}$$

is the average distance travelled at a stretch. Note that we have normalized the probabilities $W_s(X)$ such that $\sum_X [W_s(X) + W_s(-X)] = 1$.

The probability of direction reversal for the BRW is $2p(1-p)$ and has a maximum value of $f = 1/2$ at $p = 1/2$, which corresponds to a random walk. However, we get the result that, for the CCM model, f is always greater than 1/2. The data are shown in Fig. 4.19. Thus, there is no way one can extract an equivalent value of p and make comparisons. This again shows that the agents in the CCM model do not perform a biased random walk in the gain loss space.

One can also define a quantity

$$\langle X \rangle_- = \sum_X \left[X W_s(X) - X W_s(-X) \right] \tag{4.29}$$

to obtain an effective p value for each λ using the fact that, for the BRW, $\langle X \rangle_- = (2p-1)/(2p(1-p))$. $\langle X \rangle_-$ is shown as a function of λ in Fig. 4.20. Interestingly, here it is possible to extract effective values of p that are quite close to p_{slope}, the values obtained using the slope method (data shown in Fig. 4.20 to be compared with the data in Fig. 4.17).

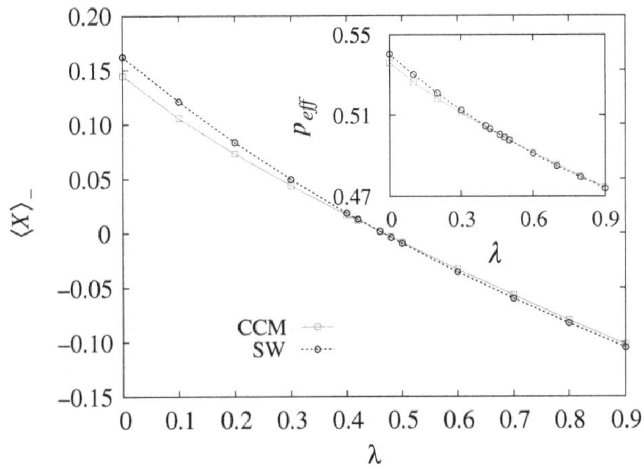

Figure 4.20 $\langle X \rangle_-$ plotted against λ for both the CCM model and SW. Inset shows the effective p values using $\langle X \rangle_-$. Reproduced from Goswami *et al.* (2011).

Thus, it is found that the results for f (which is related to $\langle X \rangle$) indicate that the CCM walk on the GLS cannot be regarded as a BRW while the measure $\langle X \rangle_-$ is fairly consistent with it. In the following subsections we resolve this intriguing issue.

It is important to make a few comments about the quantities $\langle X \rangle$ and $\langle X \rangle_-$, the first of which is directly related to the direction reversal probability f. If the left and right moves of a walker are regarded as the states of an Ising spin and the temporal sequence of the moves are viewed as spin states of the consecutive sites of a one-dimensional lattice, then $\langle X \rangle$ is equivalent to the average domain size and $\langle X \rangle_-$ can be interpreted as the magnetization. Also, it should be noted that $\langle X \rangle_-$ is a quantity which will be zero if the $W_s(X)$ distribution is symmetric.

4.1.5.4 Correlations

Earlier, it had been mentioned that, for the CC model, the walkers apparently behave as ordinary random walkers. Since the probability of direction reversal in the CCM model shows drastic difference when compared with the BRW, the question whether f is exactly equal to $1/2$ in the CC model (as in a random walk) can also be raised. Interestingly, the CC model shows little difference from the CCM when f is compared (Fig. 4.19), while previous studies had shown that the scaling of $\langle x^2 \rangle$ is quite different for the two models. We are therefore led to investigate more into the details of the walks for both the CC and CCM walks in the context of direction changes. In the CCM model, $f > 1/2$ while a bias depending on λ is simultaneously maintained. It may seem a little difficult to conceive such a walk, but it is possible to construct some deterministic toy walk models which have these properties. For example, a walk which goes along right (R) and left (L) as RRLRRL, etc. has these features. Here there is an overall bias towards right while $f = 2/3 > 1/2$. Adding some noise may still maintain the bias and $f > 1/2$. The CC walkers on the other hand also show a deviation from a simple random walk as $f > 1/2$ is obtained here.

Since a large value of probability of direction changes implies that there is a higher probability of taking two successive steps in directions opposite to each other, it immediately suggests that there is a correlation between successive steps. Let the step taken at time t be written as $s(t) = \pm 1$ ($+1$ for a right step and -1 for a left step). The time correlation function $C(t)$ is then defined as

$$C(t) = \langle s(t_0)s(t_0 + t)\rangle - \langle s(t_0)\rangle\langle s(t_0 + t)\rangle, \tag{4.30}$$

where t_0 is an arbitrary time after equilibrium is reached. The average over different initial times t_0 was taken to calculate the above correlation in a single realization of a walk for both the CC and CCM models. The second term on the right-hand side of Eq. 4.30 can be replaced by ${s_0}^2$, as $\langle s(t_0)\rangle$, the average step length, is independent of time at equilibrium and equivalent to s_0, the slope of the $\langle x\rangle$ versus t plot. For the CC walk, therefore, $\langle s(t_0)\rangle = 0$, while for the CCM walk it has a non-zero value. We notice that for both CC and CCM walks, there is a strong correlation when $t = 1$, which decays quite fast for both models. For the CC walk, the correlations become zero at later times (Fig. 4.21). For the CCM model, however, the correlation saturates to a very small non-zero value which is λ dependent. The saturation value $\bar{C} = C(t \to \infty)$ is estimated by averaging $C(t)$ over the last few hundred steps. The average saturation values $\bar{C}$ are shown in the inset of Fig. 4.21 as a function of λ. $\bar{C}$ has a minimum value $\sim O(10^{-5})$ close to λ^* and a small positive value which increases as λ deviates from λ^*.

The short time correlation in both models is indeed negative, which is consistent with the fact that direction reversal occurs with a probability greater than 1/2. It may be mentioned that, for a RW as well as a BRW, all time correlations are simply zero.

In a one-dimensional walk, two successive steps give rise to four possible paths: LR, LL, RL, RR. The probabilities of these moves were investigated in detail to gain further insight into the walks in the GLS as the correlations for successive time steps are strongest. This correlation, $C(1)$, is related to the probabilities W of these moves; precisely, $C(1) = W(RR) + W(LL) - W(LR) - W(RL) - s_0^2$.

The results for both CC and CCM models are shown in Fig. 4.22. It can be noted that, irrespective of the value of λ, $W(RL) = W(LR)$, i.e. the tendency to change direction does not depend on the sequence of the steps taken. At the same time, we note that while, for the CC walkers, there is also a symmetry $W(RR) = W(LL)$, for the CCM walkers, which have a bias, these two measures are unequal in general

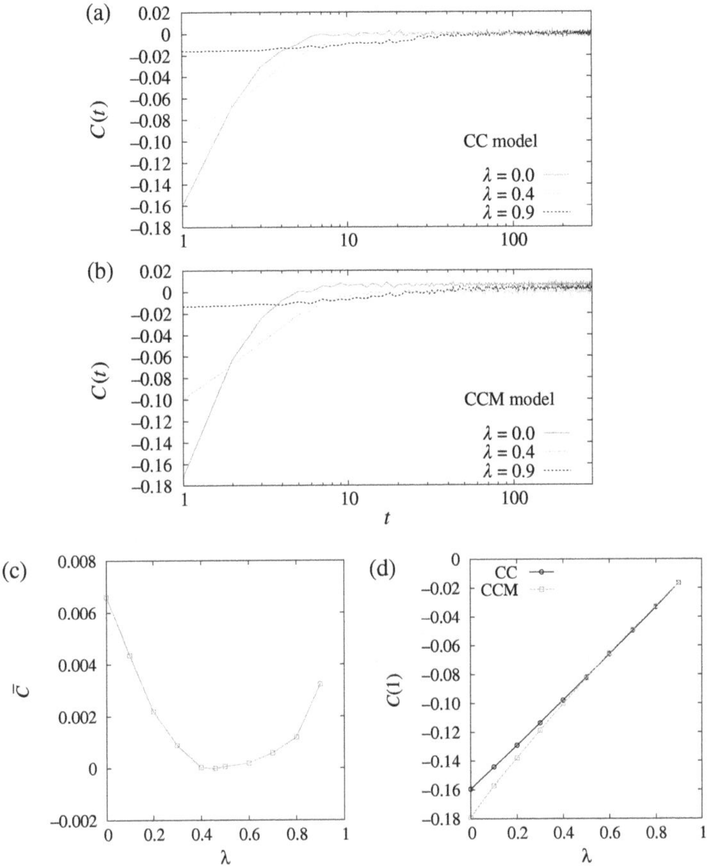

Figure 4.21 (a) The correlation of steps taken at time intervals of t is calculated using Eq. 4.30 for a single walk configuration for the CC model, averaging over all possible initial times t_0. (b) Same for the CCM model. Inset shows that the saturation value of the correlation, $\bar{C}$, at long times has a dependence on λ; it is ≈ 0 at λ^* and increases as λ deviates from λ^*. (c) Saturation value of the correlation $\bar{C}$, at long times for the CCM model shows a dependence on λ; it is ≈ 0 at λ^* and increases as λ deviates from λ^*. In (d), the correlation for two consecutive time steps, $C(1)$, is shown for both the CC and CCM models also as a function of λ. Reproduced from Goswami *et al.* (2011).

and become equal only at the 'bias-less' point λ^*. From these detailed measures, it is now entirely clear how the CC walk differs from the RW and the CCM from the BRW as illustrated in Fig. 4.23.

Understanding why direction change is preferred At this point it is apparent that, in general in these kinetic exchange models, the tendency to make a gain and a loss in successive steps (in either order) is independent of the saving feature of

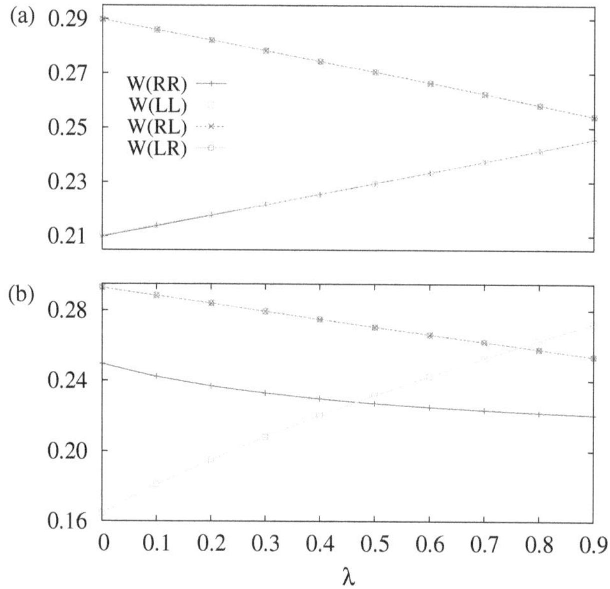

Figure 4.22 The probabilities W(RR), W(LL), W(RL) and W(LR) are shown for (a) the CC and (b) the CCM models. Reproduced from Goswami *et al.* (2011).

the CC and CCM models. In fact, it is present with maximum probability in the CC model with $\lambda = 0$ (i.e. the DY model) when agents do not save at all.

One can therefore try to understand this feature from the point of view of the DY model, which has a simple, exactly known form for the money distribution by considering the transactions made in two successive steps. It could be shown that indeed, for the DY case, it can be proved that the probability of direction changes is greater than $\frac{1}{2}$.

In the DY model (and in fact in the CC model for any λ), in general, an agent gains/loses while interacting with a richer/poorer agent. This is because if agent 1 with money m_1 interacts with agent 2 with money m_2, after interaction, agent 1 will have money $m' = \epsilon(m_1 + m_2)$. On an average, if agent 1 gains, $(m_1 + m_2)/2 > m_1$, or $m_2 > m_1$. To prove that the probability of direction changes is greater than $\frac{1}{2}$, we show that individually $W(RL)$ and $W(LR)$ are greater than 1/4. Suppose an agent had a gain in the first step and ended up with money m_g. Let $W'(LR)$ be the conditional probability that the agent loses in the next step while interacting with another agent with money m, given that she/he gained in the first step. This probability has to take care of two factors:

(1) condition that $m \leq m_g$;
(2) averaging over all possible m_g.

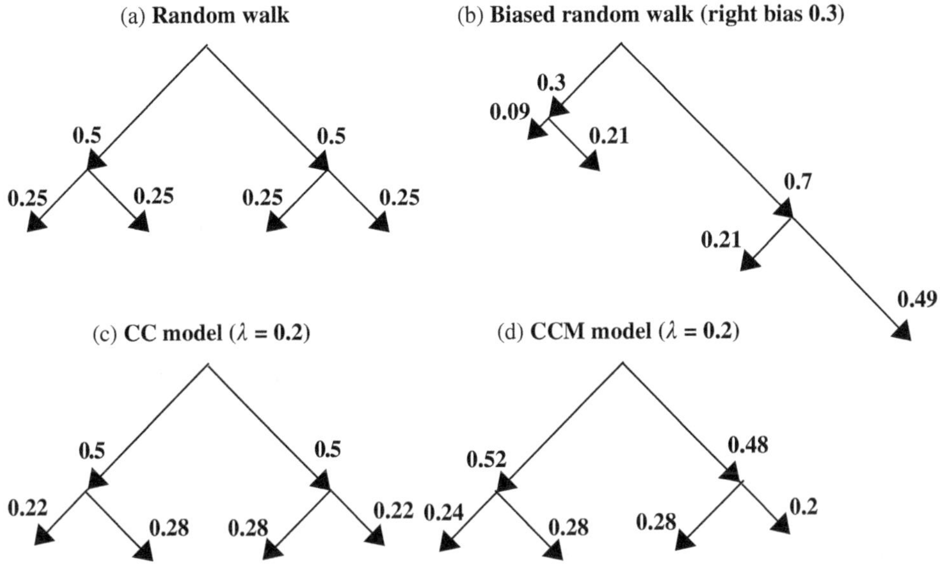

Figure 4.23 Typical movements of a walker in two successive time steps. The four possible moves – LL, LR, RL, RR – are shown with the step lengths proportional to the corresponding probabilities (also shown). In all cases, the probability of a LR move is equal to a RL move (symmetric). In (a), the movement of a random walker shows that it moves with full symmetry. (b) A biased random walker has equal probabilities for LR and RL but unequal probabilities for the other two moves. (c) CC model: walkers have equal probabilities to move L and R in step 1 like random walkers, but different probabilities for LR and LL (or RR) in the next step. However, there is symmetry for the RR and LL moves here. (d) For the CCM model, the symmetry properties are similar to the biased random walker but to be noted is the fact that LR and RL moves occur with probability >0.25 (also true for CC walks). Results are shown for $\lambda = 0.2$ for CC and CCM walks; qualitative feature is independent of λ. Reproduced from Goswami *et al.* (2011).

Using the money distribution function $P(m) = \exp(-m)$ for the DY model (taking $T = M/N = 1$), one gets

$$W'(LR) = \frac{\int_{m_g^0}^{\infty} P(m_g)dm_g \int_0^{m_g} P(m)dm}{\int_{m_g^0}^{\infty} P(m_g)dm_g} = 1 - \frac{1}{2}\exp\left(-m_g^0\right). \qquad (4.31)$$

The lower limit of the integral over m_g is taken as m_g^0 and not zero since, after a gain in the first step, the agent must have money greater than zero. m_g^0 may be considered to be an arbitrary lower bound.

Now $W(LR) = W'(LR)/2$ simply as we know that probability of R or L at the first step is just 1/2 (Chatterjee and Sen 2010). Therefore one would get $W(LR) = [1 - \frac{1}{2}\exp(-m_g^0)]/2 \geq 1/4$ independent of the value of m_g^0.

In a similar manner, for the move RL, one has an arbitrary upper bound m_l^0 for the money m_l the agent ends up with after a loss in the first step such that

$$W(RL) = \frac{1}{2}\frac{\int_0^{m_l^0} P(m_l)dm_l \int_{m_l}^{\infty} P(m)dm}{\int_0^{m_l^0} P(m_l)dm_l} = \frac{1}{4}[1 + \exp(-m_l^0)]. \quad (4.32)$$

Obviously, $W(RL)$ is also greater than or equal to 1/4 for all values of m_l^0 and therefore the sum $W(LR) + W(RL) \geq \frac{1}{2}$. Since $W(RL)$ and $W(LR)$ equal 1/4 for extremely improbable cases, one can conclude that $W(LR) + W(RL) > \frac{1}{2}$ in general.

Now, what happens for the CC and CCM models? In the CC model, the conditional probability that an agent loses after gaining depends on λ through the money distribution function. Since its form is not exactly known, it is not possible to get exact results. However, for the CC model, there is a growing region for the money distribution curve for small m values and therefore the probability that agent 1 meets a poorer agent in the next step is less probable than in the DY model and hence qualitatively it is understandable that $W(RL)$ or $(W(LR))$ will decrease with λ. At the same time it is true in the CC model also that the conditional probability W' is twice the probability W of a LR or a RL move as in the DY model, independent of λ (Chatterjee and Sen 2010).

In the CCM model, matters become more complicated as the condition for gain/loss depends on the interacting agents' saving propensities. It was found in Chatterjee and Sen (2010) that the probability of a gain is higher when one interacts with an agent with larger λ. Since the average money of an agent increases with λ (Mohanty 2006) (in a non-linear manner), this condition implies, once again, that a gain is more likely while interacting with a richer agent. Consequently, the same kind of logic holds good here: for a direction change to occur, an agent who got richer (poorer) in one step should interact with a poorer (richer) agent in the next. However, like the CC model, the exact form of the money distribution is not known here. Moreover, the probabilities W' and W for the LR and RL moves are not simply related in the CCM model.

4.1.5.5 Comparisons with a simulated walk of a single agent

In the previous section it was concluded that the form of the money distribution is responsible for the preference of direction change in the gain–loss space, although the actual amount of money lost/gained is ignored in the walk picture. In fact, in these types of kinetic exchange models, whether there is saving or not, the choice of the second agent becomes an important factor, giving rise to $f > 1/2$ in all cases. Hence one is led to believe that, if a CCM/CC kind of walk is generated which does not take into account the above choice, the result $f > 1/2$ will not be observed. Such a walk for the CC is trivial, one only has to generate a walk which has probability $1/2$ of going either way making it completely identical to a RW.

For the CCM, however, it is possible to generate a non-trivial single-agent walk in which the choice of the second agent is arbitrary. It was found in Chatterjee and Sen (2010) that the probability of gain over loss on an average for an agent with given saving propensity λ_1, while interacting with another agent whose saving propensity is λ_2, has the following form (4.23):

$$\mathcal{P}_g - \mathcal{P}_l = \text{const.} \frac{\lambda_2 - \lambda_1}{1.5 + \lambda_1 + \lambda_2}.$$

The constant turns out to be very close to 0.345. This equation suggests that, at each step, a tagged walker with saving λ_1 will move left/right with a probability which depends on λ_2 as well. This probability at each step can be calculated easily from the above equation once λ_1 and λ_2 values are known and using the fact that $\mathcal{P}_g + \mathcal{P}_l = 1$. It is therefore possible to generate a walk for a single agent with given λ_1, assuming that at each step it interacts with another agent of *randomly chosen* λ_2 to give the probability of movement to right/left at that instant. Thus, in this walk, the choice of λ_2 is completely random; the money distribution function does not enter the picture at all and at the same time the probability of a move towards any direction is not fixed.

It is interesting to compare the results of this simulated walk with the original multiagent CCM walk. We find that in fact the effective p values are almost identical. For the simulated walk (SW), we can extract an effective p in two ways: first is as usual by calculating the slope (section IIA), and secondly by taking the average value of the probability $\mathcal{P}_g$ (to move right) generated for all times – these two values are very close. Only the value obtained using the slopes have been shown in Fig 4.17 along with the results for the CCM walk.

However, when one calculates $W_s(X)$ for the simulated walks, it turns out that these are not at all comparable to the CCM (Fig 4.24). There are two interesting features to be noted here: the probabilities for small X are larger for the CCM model and the magnitude of differences decrease with λ. Both these results can be explained from the anti-persistence effect present in the CCM model. Here the increased number of direction changes results in a larger value of $W_s(X)$ for small X and the fact that the anti-persistence effect decreases with λ makes the CCM and SW models more similar as λ increases. It was also noted that the p values extracted from the ratio $W_s(X)/W_s(-X)$ are indeed independent of λ (Fig. 4.25), which is expected for a BRW. So the simulated single agent walk is like a conventional BRW, compared with the CCM, in which the p values have a dependence on λ as well as on the number of steps X (Fig. 4.18).

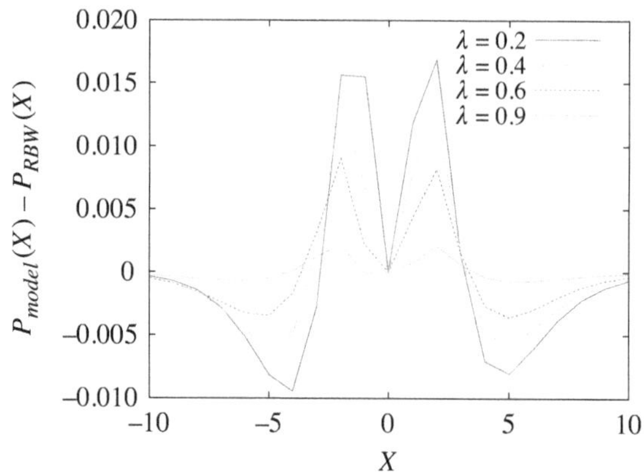

Figure 4.24 Difference between the probabilities $W_s(X)$ calculated from the CCM model and the simulated walk are plotted against λ. Reproduced from Goswami *et al.* (2011).

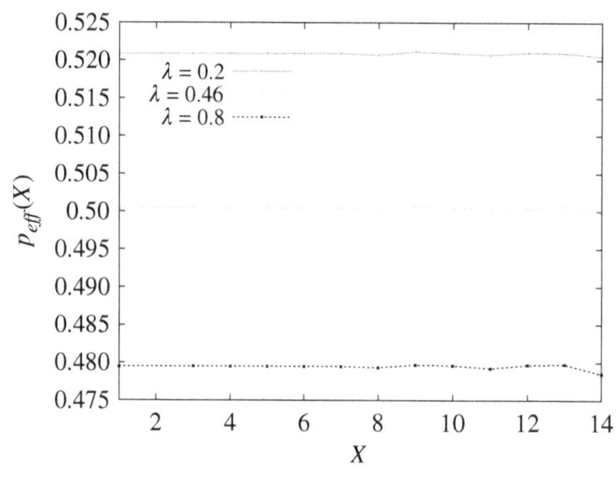

Figure 4.25 Variation of $p_{eff}(X)$ against X for different $\lambda = 0.0, 0.2, 0.46, 0.8$ for the simulated walk. Reproduced from Goswami *et al.* (2011).

When f, the fraction of direction changes, is calculated for the simulated walk, we find that it is less than $\frac{1}{2}$ for all values of λ, and very close to $\frac{1}{2}$ at $\lambda^* \simeq 0.469$ (Fig. 4.19).

The simulated walk of a single agent once again shows the presence of a $\lambda^* \approx 0.47$ where the walk becomes bias-less, but otherwise shows features which are identical to those of BRW. This is consistent with the conjecture that the choice of the second agent is crucial when the money distribution form plays a significant role.

4.2 Models with commodity

4.2.1 Ideal-gas trading market in the presence of a non-consumable commodity

In the above markets, modifications owing to exchange of a consumable commodity hardy affects the distribution, as the commodity once bought or sold need not be accounted for. Consumable commodities effectively have no 'price', because of their short lifetime to contribute to the total wealth of an individual. It is interesting, however, to study the role of non-consumable commodities in such market models, and this we do here.

In a simplified version of a market with a single non-consumable commodity, Chatterjee and Chakrabarti (2006) considered a fixed number of traders or agents N who trade in a market involving total money $\sum_i m_i(t) = M$ and total commodity $\sum_i c_i(t) = C$, $m_i(t)$ and $c_i(t)$ being the money and commodity of the i-th agent at time t and both are non-negative. Needless to mention, both $m_i(t)$ and $c_i(t)$ change with time or trading t. The market, as seen before, is closed, which means that N, M and C are constants. The wealth w_i of an individual i is thus the sum of the money and commodity it possesses, i.e. $w_i = m_i + p_0 c_i$; p_0 is the 'global' price. In the course of trading, money and commodity are locally conserved, and hence the total wealth. In such a market, one can define a global average price parameter $p_0 = M/C$, which is set here to unity, giving $w_i = m_i + c_i$. It may be noted that, in order to avoid the complication of restricting the commodity–money exchange and their reversal between the same agents, the Fisher velocity of money circulation (e.g. Wang *et al.* 2006) is renormalized to unity here. In order to accommodate the lack of proper information and the ability of the agents to bargain etc., one can allow of course fluctuations δ in the price of the commodities at any trading (time): $p(t) = p_0 \pm \delta = 1 \pm \delta$. It is found that the nature of steady state is unchanged and independent of δ, once it becomes non-vanishing.

4.2.1.1 Dynamics

In general, the dynamics of money in this market looks the same as Eq. (4.2), with Δm given by Eqs. (4.4), (4.9) or (4.15) depending on whether $\lambda_i = 0$ for all, $\lambda_i \neq 0$ but uniform for all i or $\lambda_i \neq \lambda_j$, respectively. However, all Δm are not allowed here; only those for which $\Delta m_i \equiv m_i(t+1) - m_i(t)$ or Δm_j are allowed by the corresponding changes Δc_i or Δc_j in their respective commodities ($\Delta m > 0$, $\Delta c > 0$):

$$c_i(t+1) = c_i(t) + \frac{m_i(t+1) - m_i(t)}{p(t)}, \tag{4.33}$$

$$c_j(t+1) = c_j(t) - \frac{m_j(t+1) - m_j(t)}{p(t)}, \tag{4.34}$$

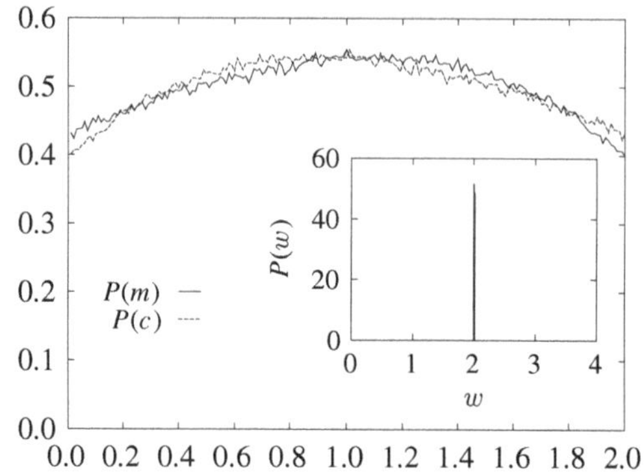

Figure 4.26 Steady-state distributions of money ($P(m)$ vs. m) and commodity ($P(c)$ vs. c) in a market with no savings (saving factor $\lambda = 0$) for no price fluctuations i.e. $\delta = 0$. The graphs show simulation results for a system of $N = 100$ agents, $M/N = 1$, $C/N = 1$; $m_i = 1 = c_i$ at $t = 0$ for all agents i. The inset shows the distribution $P(w)$ of total wealth $w = m + c$. As $p = 1$, for $\delta = 0$, although m and c can change with tradings within the limit $(0 - 2)$ the sum is always maintained at 2. Reproduced from Chatterjee and Chakrabarti (2006).

where $p(t)$ is the local-time 'price' parameter, a stochastic variable:

$$p(t) = \begin{cases} 1 + \delta & \text{with probability } 0.5, \\ 1 - \delta & \text{with probability } 0.5. \end{cases} \tag{4.35}$$

The role of the stochasticity in $p(t)$ is to imitate the effect of bargaining in a trading process, δ parametrizes the amount of stochasticity. The role of δ is significant in the sense that it determines the (relaxation) time the whole system takes to reach a steady state; the system reaches equilibrium sooner for larger δ, while its magnitude does not affect the steady-state distribution. It may be noted that, in the course of the trading process, certain exchanges are not allowed (e.g. in cases when a particular pair of traders do not have enough commodity to exchange in favour of an agreed exchange of money). One then skips these steps and chooses a new pair of agents for trading.

4.2.1.2 Results

For $\delta = 0$, of course, the wealth of each agent remains invariant with time; only the proportion of money and commodity interchange within themselves, since the 'price' factor remains constant. This of course happens irrespective of the savings factor being zero, uniform or distributed. For $\delta = 0$, the steady-state distribution of money or commodity can take non-trivial forms (Fig. 4.26), but has strictly a δ-function behaviour for the total wealth distribution; it gets frozen at the value of wealth one starts with (see inset of Fig. 4.26 for the case $m_i = 1 = c_i$ for all i).

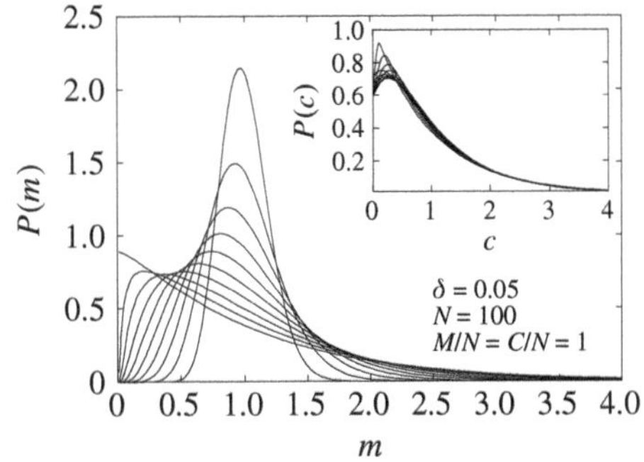

Figure 4.27 Steady-state distribution $P(m)$ of money m in the uniform savings commodity market for different values of saving factor λ $(0, 0.1, 0.2, 0.3, 0.4, 0.5, 0.6, 0.7, 0.8, 0.9$ from left to right near the origin) for $\delta = 0.05$. The inset shows the distribution $P(c)$ of commodity c in the uniform savings commodity market for different values of saving factor λ. The plots show simulation results for a system of $N = 100$ agents, $M/N = 1$, $C/N = 1$. Reproduced from Chatterjee and Chakrabarti (2006).

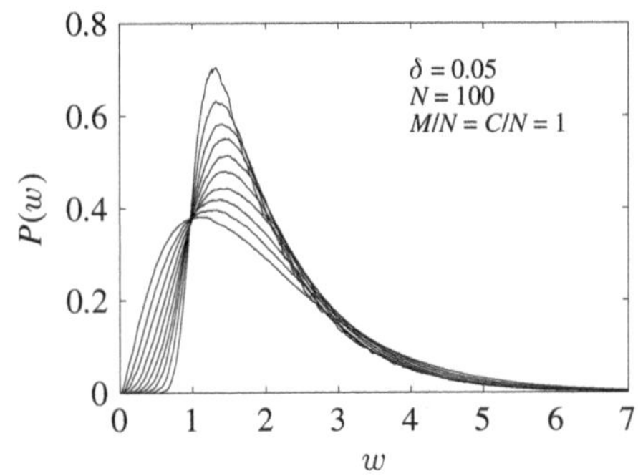

Figure 4.28 Steady-state distribution $P(w)$ of total wealth $w = m + c$ in the uniform savings commodity market for different values of saving factor λ $(0, 0.1, 0.2, 0.3, 0.4, 0.5, 0.6, 0.7, 0.8, 0.9$ from left to right) for $\delta = 0.05$. The plots show simulation results for a system of $N = 100$ agents, $M/N = 1$, $C/N = 1$. Reproduced from Chatterjee and Chakrabarti (2006).

As mentioned already for $\delta \neq 0$, the steady-state results are not dependent on the value of δ (the relaxation time of course decreases with increasing δ). In such a market with uniform savings, money distribution $P(m)$ has a form similar to a set (for $\lambda \neq 0$) of gamma functions (Fig. 4.27): a set of curves with a most-probable value shifting towards 1 from below, as the saving factor λ changes from 0 to 1 (as in the case without any commodity). The commodity distribution $P(c)$ has an initial peak and an exponential fall-off, without much systematics with varying λ (see inset of Fig. 4.27). The distribution $P(w)$ of total wealth $w = m + c$ behaves much like $P(m)$ (Fig. 4.28). It is to be noted that, since there is no precise correspondence

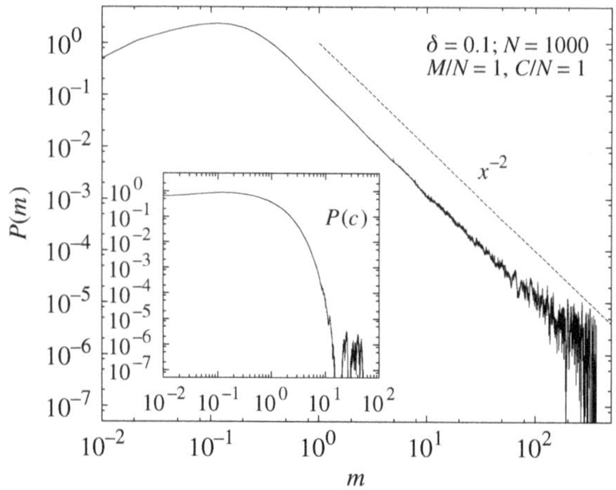

Figure 4.29 Steady-state distribution $P(m)$ of money m in the commodity market with distributed savings $0 \leq \lambda < 1$. $P(m)$ has a power law tail with Pareto index $\nu = 1.00 \pm 0.02$ (a power law function x^{-2} is given for comparison). The inset shows the distribution $P(c)$ of commodity c in the same commodity market. The plots show simulation results for a system of $N = 1000$ agents, $M/N = 1$, $C/N = 1$. Reproduced from Chatterjee and Chakrabarti (2006).

with commodity and money for $\delta \neq 0$ (unlike when $\delta = 0$, when the sum is fixed), $P(w)$ cannot be derived directly from $P(m)$ and $P(c)$. However, there are further interesting features. Although they closely resemble gamma distributions, the set of curves for different values of saving factor λ seem to intersect at a common point, near $w = 1$, which might reveal a deeper meaning on further investigation. All the reported data are for a system of $N = 100$ agents, with $M/N = 1$ and $C/N = 1$ and for a case when the noise level δ equals 10%.

For λ distributed uniformly within the interval $0 \leq \lambda < 1$, the tails of both money and wealth distributions $P(m)$ and $P(w)$ have Pareto law behaviour with a fitting exponent value $\nu = 1.00 \pm 0.02$ and $\nu = 1.00 \pm 0.05$, respectively (Figs. 4.29 and 4.30, respectively), whereas the commodity distribution is still exponentially decaying (see inset of Fig. 4.29).

Hence, the presence of a non-consumable commodity can affect the nature of money and wealth distributions. However, under a kinetic exchange dynamics of the CCM form, the resultant distribution of money and wealth shows the self-organizing effects and retains the power law forms (Chatterjee and Chakrabarti 2006).

4.2.2 Self-organizing model with single commodity

Adam Smith in 1776 first considered the self-organizing aspect of a market consisting of selfish agents, which he called the 'invisible hand' effect (Smith 1776). The

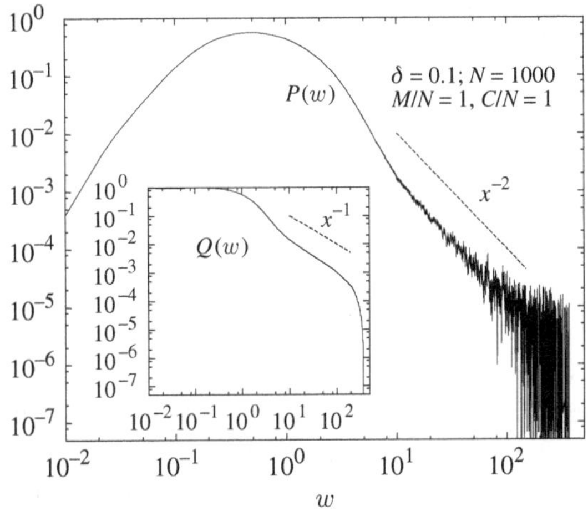

Figure 4.30 Steady-state distribution $P(w)$ of total wealth $w = m + c$ in the commodity market with distributed savings $0 \leq \lambda < 1$. $P(w)$ has a power law tail with Pareto index $\nu = 1.00 \pm 0.05$ (a power law function x^{-1} is given for comparison). The inset shows the cumulative distribution $Q(w) \equiv \int_w^\infty P(w)dw$. The plots show simulation results for a system of $N = 1000$ agents, $M/N = 1$, $C/N = 1$. Reproduced from Chatterjee and Chakrabarti (2006).

mainstream economists seem to consider it to be true in principle (as a matter of faith?) (Samuelson 1998). In various statistical physics models of interacting systems or networks, such self-organization has indeed been demonstrated to emerge in the global aspects of the system which consists of a large number of simple dynamical elements having local (in time and space) interactions and dynamics (Bak 1996).

Chakraborti *et al.* (2001) studied the self-organizing features of the dynamics of a model market, where the agents 'trade' for a single commodity with their money. They demonstrated that the model, apart from having a self-organizing behaviour, has got a crucial role for the money supply in the market and that its self-organizing behaviour is significantly affected when the money supply becomes less than the optimum. They also observed that this optimal money supply level of the market depends on the amount of 'frustration' or scarcity in the commodity market. In their model, each agent having commodity less than the 'subsistence' level traded with any other having more than the 'subsistence' level in exchange of its money.

Specifically, they considered a closed economic system consisting of N economic agents, where each economic agent i has at any time money m_i and commodity q_i, such that $(\sum_{i=1}^N m_i = M$ and $\sum_{i=1}^N q_i = Q)$, where N, M and Q are fixed. The 'subsistence' commodity level for each agent is q_0. Hence at any time an agent having $q_i < q_0$ will attempt to trade, utilizing its money m_i at that time, with agents having commodity more than q_0, and will purchase to make its commodity level q_0 (and no further), if its money permits. The agents with $q_i > q_0$ will sell off the excess amount to such 'hungry' agents having $q_i < q_0$, and will attempt to maximize their wealth (money). This dynamics is local in time ('daily') and it stops eventually when no further trade is possible satisfying the above criteria.

They introduced an 'annual' or long-time dynamics when some random fluctuations in all the agents' money and commodity occur. Annually, each agent would get a minor reshuffle of the money and its commodity (e.g. perhaps owing to the noise in the stock market and in the harvest because of the changes in the weather respectively). This (short- and long-time) combined dynamics is similar to that of the 'sand-pile' models (Bak 1996). In this model, the price of the commodity does not change with the money supply M in the market; it remains fixed (at unity). The aim was to find the steady-state features of this market; in particular, the distributions $P(m)$ and $P(q)$ of the money and commodity, respectively, among the agents. They investigated how many agents $P(q_0)$ can satisfy their basic needs through this dynamics, i.e. can reach the subsistence level q_0, as a function of the money supply M for both the unfrustrated ($g < 1$) (where $g = q_0 / < q >$ and $< q > = Q/N$ is the average commodity in the market) and the frustrated ($g > 1$) cases of the commodity market. Interestingly, they observed that an optimum amount M_0 of money supply is required for evolving the market towards the maximum possible value of $P(q_0)$, and this optimum value of money M_0 was observed to decrease with an increase in g in the market. This corroborates the view of the economists that: (1) this sort of dynamics, which takes the system to equilibrium, is greatly facilitated by 'paper money', which does not have any value of its own, but can be considered rather as a good 'lubricant' in the economic system (Samuelson 1998) and (2) when the (paper) money supply gets changed, it does not just scale up (for increased money supply) or down (for decreased money supply) the commodity prices, the (self-organizing) dynamics towards equilibrium also gets seriously affected (Keynes 1937).

4.2.2.1 Unlimited money supply and limited supply of commodity

First, they considered the money supply M in the market to be infinitely large, so that it dropped out from any consideration. The dynamics is then entirely governed by the commodity distribution among agents: for agents with $q_i < q_0$, the attempt will be to find another trade partner having $q_i > q_0$; and having found such partners, through a random search in the market, trades occur for mutual benefit (for the selling agent one still considers the extra money from trade to be important). The system thus evolves towards its steady state, as the fixed-point feature of the short-time or daily dynamics gets affected by the random noise reshuffling in the commodity of each agent. This reshuffling essentially induces Gibbs-like distribution (Chakraborti and Chakrabarti 2000). The trade dynamics is clearly

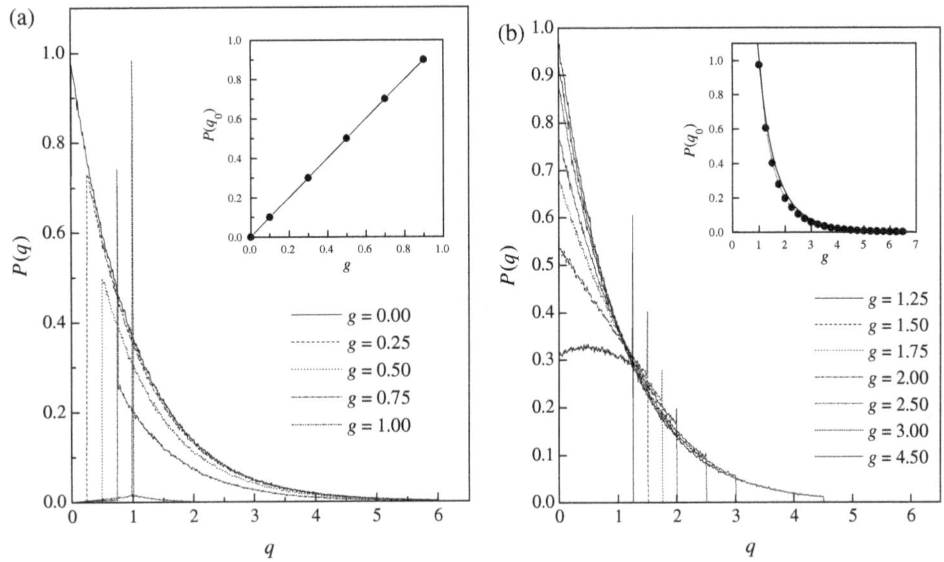

Figure 4.31 (a) The distributions of commodity $P(q)$ for different values of g for $N = 1000$, $Q = 1, M = 100\,(M > M_0(g))$, for the unfrustrated case ($g < 1$). The steady-state distribution of commodity $P(q)$ is Gibbs-like: $P(q) = A\exp(-q/ < q >)$ with $A = 1 - g$, for $q > q_0$. The inset shows the linear variation of $P(q_0)$ with g ($P(q_0) = g$). (b) The distributions of commodity $P(q)$ for different values of g for $N = 1000$, $Q = 1$, $M = 100$ ($M > M_0(g)$), for the frustrated case ($g > 1$). The variation of $P(q_0)$ with g is shown in the inset where the theoretical estimate ($P(q) = g\exp(-g)/(g - 1 + \exp(-g))$) is also indicated by the solid line. Reproduced from Chakraborti *et al.* (2001).

motivated or 'directed'. They looked for the combined effect on the steady-state distribution of commodity $P(q)$, which is independent of the initial commodity distribution among the agents.

For the unfrustrated case ($g < 1$), where *all* the agents can be satisfied, the typical distributions $P(q)$ are shown in Fig. 4.31 for different values of g. It is seen that the $P(q)$ is Gibbs-like ($P(q) = A\exp(-q/ < q >)$ and $A = 1 - g$), for $q > q_0$, while $P(q_0) = g$ (as shown in the inset). One can easily explain these observations using the fact that the cumulative effect of the long-time randomization gives Gibbs distribution ($\exp(-q/ < q >)$ for all q). They then estimated the final steady-state distribution $P(q)$ from the additional effect of the short-time dynamics on this (long-time dynamics induced) Gibbs distribution. All the agents with $q < q_0$ manage to acquire a q_0 level of commodity (as $g < 1$ and everybody has enough money to purchase the required amount). Their number is then given by $N_- = \int_0^{q_0} \exp(-q)dq$. They require the total amount of commodity $q_0 N_-$. The amount of commodity already available to them is given by $Q_- = \int_0^{q_0} q\exp(-q)dq$. The excess amount required $Q_{demand} = q_0 N_- - Q_-$ has to come from the agents having $q > q_0$. The average of the excess amount of commodity of the agents who are above the q_0 line is given by $< q_{excess} >= (1 - Q_- - (1 - N_-)q_0)/(1 - N_-)$. The number of agents who supply the Q_{demand} amount is given by $N_+ = Q_{demand}/ < q_{excess} >$. This gives $P(q_0) = N_- + N_+ = g$. One can easily determine the prefactor of the final steady-state distribution $P(q)$ for $q > q_0$, $A = 1 - g$ from the conservation of the total number of agents and total commodity.

For the frustrated case ($g > 1$), the results are shown in Fig. 4.31. A similar calculation for $P(q_0)$ is done as follows: $N_+ = \int_{q_0}^{\infty} \exp(-q)dq$ is the number of people above the q_0 line who will sell off their excess amount of commodity to come to the q_0 level, $Q_+ = \int_{q_0}^{\infty} q \exp(-q)dq$ is the commodity of the agents above the q_0 level. Then the supplied amount of commodity to the agents below the q_0 line is $Q_{supply} = Q_+ - q_0 N_+$. The average of the deficit commodity, $< q_{deficit} >= ((1 - N_+)q_0 - 1 + Q_+)/(1 - N_+)$. Hence, the number of agents who would reach the q_0 level from below is $N_- = Q_{supply}/ < q_{deficit} >$, so that $P(q_0) = N_+ + N_- = g \exp(-g)/(g - 1 + \exp(-g))$. A comparison of this estimate for $P(q_0)$ with g is also shown in the inset of Fig. 4.31. It may be mentioned that, in absence of the strict Gibbs distribution for $P(q)$ ($q < q_0$), the above expression for $P(q_0)$ is somewhat approximate.

4.2.2.2 Limited money supply and limited supply of commodity

When the money supply is limited, the self-organizing behaviour is significantly affected and the fraction of agents who can secure q_0 amount of commodity for themselves $P(q_0)$ does not always reach its maximum possible value (as suggested by the amount of commodity available in the market). For all values of g, as one increases the money supply in the market M, $P(q_0)$ increases and then after a certain amount M_0, it saturates. In Fig. 4.32, it has been shown how the quantity $P(q_0)$ varies with M for different values of g (for $g > 1$ only, as one is more interested in the frustrated case). One defines M_0 to be the optimum amount of money supply needed for the smooth functioning of the market. They also observed that this optimal money supply level of the market depends on the amount of frustration g in the commodity market. In the inset, the variation of M_0 with g is shown for the frustrated case ($g > 1$) only.

4.2.3 Another model with commodity

In another attempt to realize an economy which could be more 'realistic', Ausloos and Pękalski (2007) proposed another model that included exchangeable commodities. Similar to the models discussed in the previous subsections, it is a fully conservative model. The model consists of N trading agents with M discrete commodities and money amount M (integer) per agent (although money is continuous).

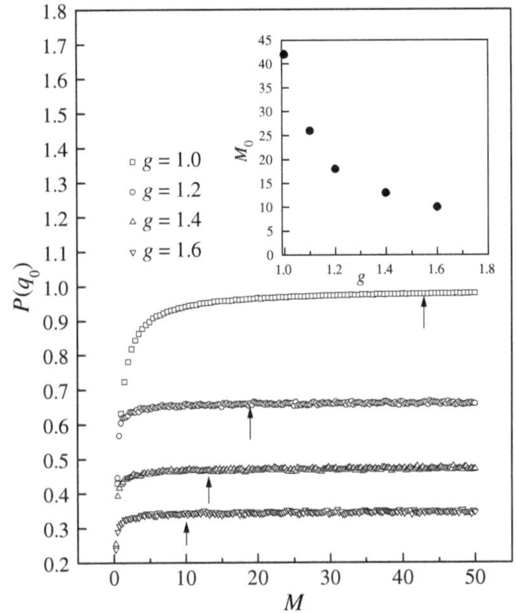

Figure 4.32 The variation of quantity $P(q_0)$ with M for different values of g in the frustrated cases ($g > 1$). The inset shows the variation of M_0 with g. Reproduced from Chakraborti *et al.* (2001).

In the model anyone is allowed to trade with anyone else provided they possess enough money or commodity to carry out an exchange. The model dynamics is defined as follows: a random agent j is chosen, and, if she has money, she chooses to buy with probability $1/2$. A random fraction of the total money is considered and the maximum amount of commodities she can buy is also calculated, which is predecided not to exceed some M_m. A second agent k is randomly chosen, and checked if she has enough commodities to sell to j. If this is satisfied, the decision to sell is taken with probability $1/2$. Next, the exchange takes place – money passes from j to k and commodities from k to j. N choices of agent j completes one Monte Carlo time step.

The first quantity of interest to the authors is the time evolution of the amount of money for the richest trader and the amount of commodity held by the trader with the maximum number of them. The curves look asymptotically flat, but the richest agent attains the maximum long before the agent with maximum number of commodities (Fig. 4.33). The authors attribute this to the fact that they restricted the upper limit of the number of commodities exchanged in a single transaction to some M_m. On the other hand, there was no a priori constraint on the price and was restricted by the amount of money of the buying agent. Despite the fact that the price is randomly determined by the seller, and the buyer also accepts or rejects it randomly, the

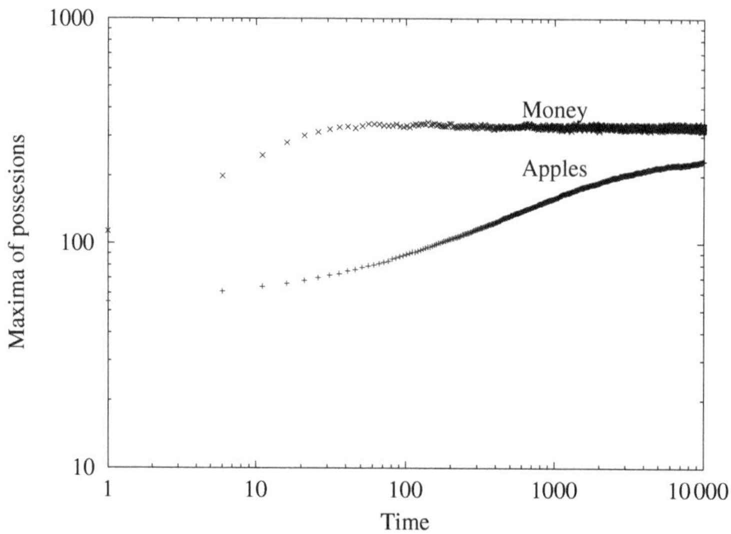

Figure 4.33 The time evolution of the maximum money possessed by the richest agent, and also the agent with maximum number of commodities, in the specific example with $N = 60$, $M = 50$, $M_m = 5$. Reproduced from Ausloos and Pękalski (2007).

number of transactions, number of exchange commodities and amount of money exchanged show little fluctuation, maintaining approximately constant throughout. The distribution of money spreads quickly and in the long time there is a very broad distribution with some agents with large amounts of money. In comparison, the distribution of commodities is not that broad, which the authors attribute to the 'socialist' restriction on the maximum number of commodities exchanged in each transaction. In their specific example (with apples as commodities) with $N = 60$, $M = 50$, $M_m = 5$ in Fig. 4.34, the authors claim two power law regions in the money distribution, the distinct regions which have been produced owing to the nature of the model and its constraints. The authors also considered the effect of taxes on their model. They defined a taxation process in which a certain fraction of money that is exchanged in each transaction goes to the 'state' and thus disappears, leaving the system 'open' with respect to money. The money evolution is thus non-monotonic for the richest agent (Fig. 4.35), while commodities remain unaffected. The overall effect on the 'society' is felt by the fact that there is a considerable change in the distribution of money in the system – the dispersion between the richest and the poorest agents is less for a society with tax, than one which has no tax (Fig. 4.36), while the distribution of commodity remains fairly unchanged. The average price per commodity goes down considerably (Fig. 4.35), while the number of poor agents increases (Fig. 4.36).

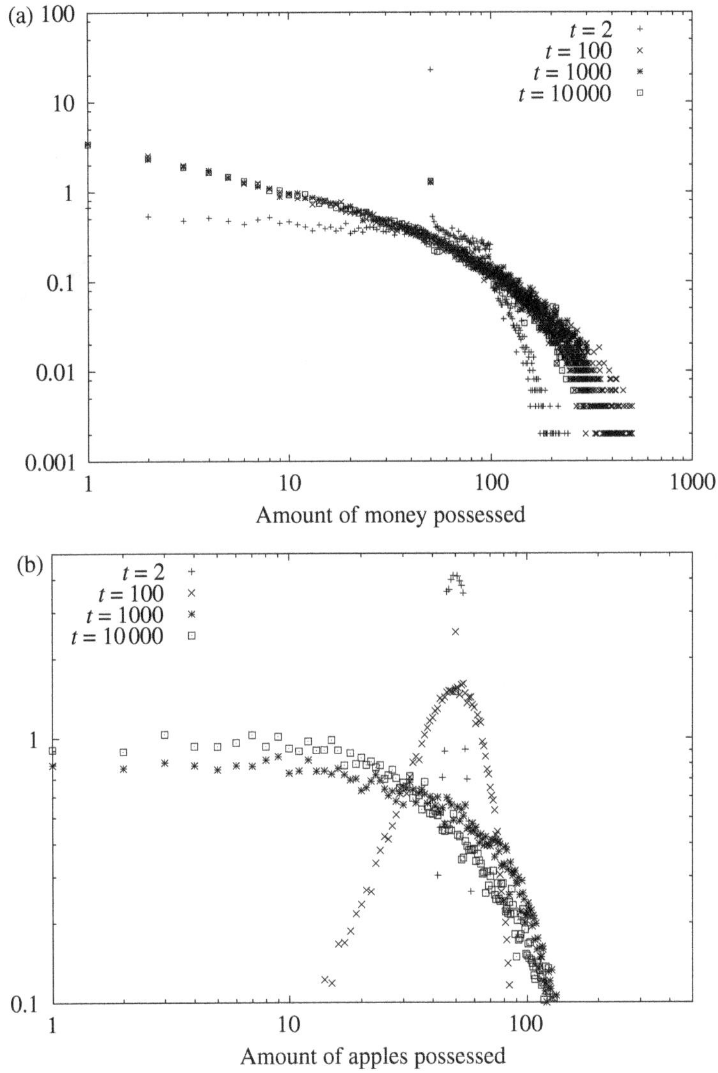

Figure 4.34 The time evolution of the distribution of money (a), and also the distribution of commodities (b), in the specific example with $N = 60$, $M = 50$, $M_m = 5$. Reproduced from Ausloos and Pękalski (2007).

4.3 Models on networks

4.3.1 *Models on directed networks*

The topology of exchange space in a real society is quite complicated. There is not only a strong sense of directionality but often a hierarchy in the underlying network – money is preferentially transferred in one direction, which results in the irreversible flow of money. A mean-field scenario (Section 4.1) does not include

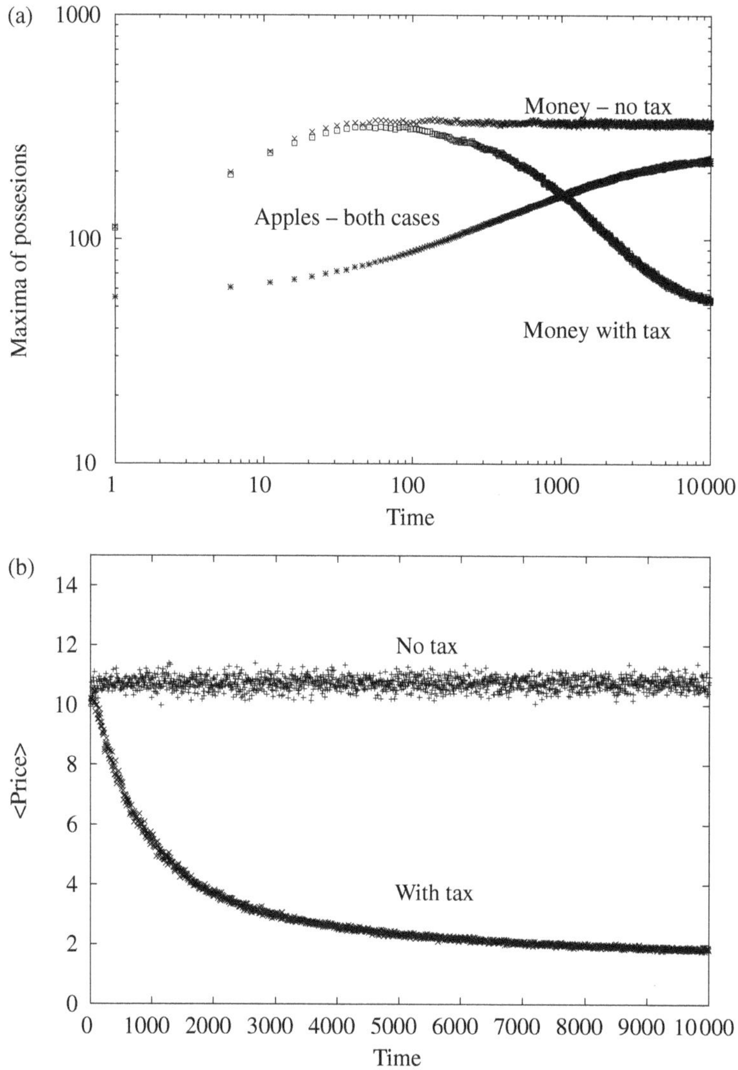

Figure 4.35 (a) The time evolution of the money and commodities with 1% tax and without tax. (b) Time evolution of the average price per unit commodity without and with tax. The specific example is with $N = 60$, $M = 50$, $M_m = 5$. Reproduced from Ausloos and Pękalski 2007.

the constraints on the flow of money or wealth. A way to imitate this is to consider wealth exchange models on a directed network (Wasserman and Faust 1994; Albert and Barabási 2002; Dorogovtsev and Mendes 2003b). There have been previous attempts to obtain the same using the physics of networks (Hu *et al.* 2006, 2007; Garlaschelli and Loffredo 2008).

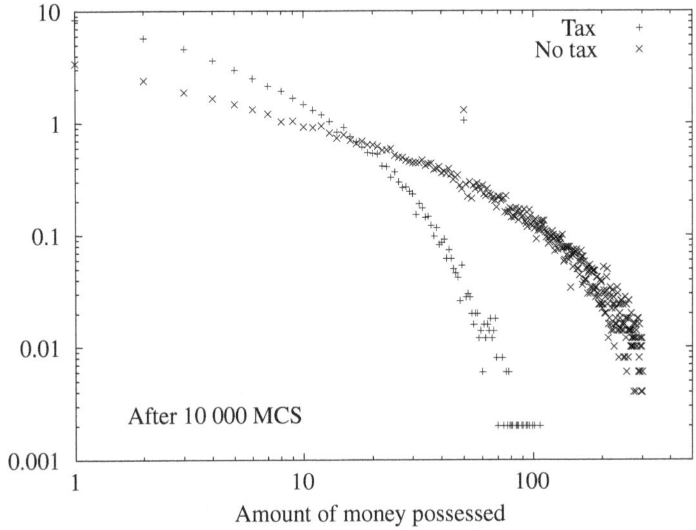

Figure 4.36 The distribution of money with 1% tax and without tax. The specific example is with $N = 60$, $M = 50$, $M_m = 5$. Reproduced from Ausloos and Pękalski (2007). MCS, Monte Carlo time step.

We consider that there are N agents, each represented by a node and connected to the rest $N - 1$ by directed links. The directionality of the links denote the direction of flow of wealth in this fully connected network. The directed network parametrized by p is constructed in the following way (Chatterjee 2009, 2010):

(1) There are no self-links, so that the adjacency matrix $\mathcal{A}$ (Albert and Barabási 2002; Dorogovtsev and Mendes 2003b) has diagonal elements $a_{ii} = 0$ for all sites i.
(2) For each matrix element a_{ij}, $i \neq j$, we call a random number $r \in [0, 1]$. $a_{ij} = +1$ if $r < p$ and $a_{ij} = -1$ otherwise. Also $a_{ji} = -a_{ij}$. Thus, we have $N(N-1)/2$ such calls of r. $a_{ij} = +1$ denotes a directed link from i to j and $a_{ij} = -1$ denotes a directed link from j to i.

The link disorder at site i is $\rho_i = \frac{1}{N-1} \sum_j a_{ij}$. $\sum_j$ denotes the sum over all $N - 1$ sites j linked to i. Thus, $\rho_i = 1$ is a node which has all links outgoing and $\rho_i = -1$ is a node for which all links are incoming. The parameter p has a symmetry about 0.5 and the distribution $R(\rho_i)$ is also symmetric about 0, which is, in fact, a consequence of the conservation of the number of incoming and outgoing links. A network with $p = 0.5$ has the lowest degree of disorder, given by a narrow distribution $R(\rho)$ of ρ, around $\rho = 0$ (Fig. 4.37). This means that almost all nodes have a more or less equal number of incoming and outgoing links. On the other extreme, $p = 0.01$ is a network which has a small but finite

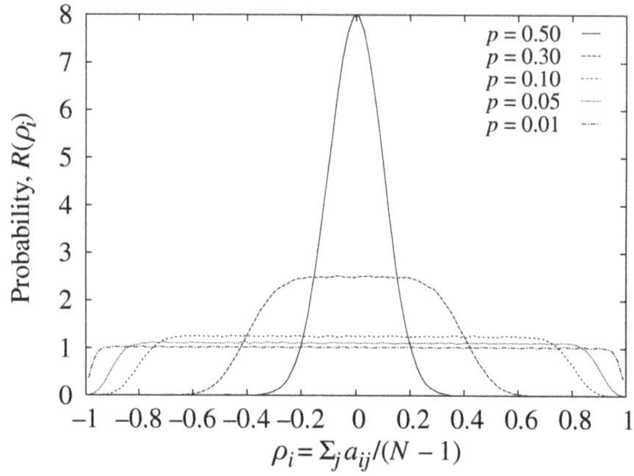

Figure 4.37 The distribution $R(\rho_i)$ of the link disorder ρ_i for a directed network with different values of $p = 0.50, 0.30, 0.10, 0.05, 0.01$ for a system of $N = 100$ nodes, obtained by numerical simulation, averaged over 10^4 realizations. Reproduced from Chatterjee (2009).

number of nodes where most links are incoming/outgoing, thus giving rise to a very wide distribution of link disorder $R(\rho)$.

The rules of exchange are: if $a_{ij} = -1$,

$$m_i(t+1) = m_i(t) + \mu_j m_j(t),$$
$$m_j(t+1) = m_j(t) - \mu_j m_j(t),$$

else, $a_{ij} = +1$,

$$m_i(t+1) = m_i(t) - \mu_i m_i(t),$$
$$m_j(t+1) = m_j(t) + \mu_i m_i(t).$$

$0 < \mu_i < 1$ is the 'transfer fraction' associated with agent i, and $m_i(t)$ is the money of agent i (or, money at node i) at time t. The total money in the system is conserved, no money is created or destroyed, defined by the above equations. If there is a link from j to i, the node i gains μ_j fraction of jth agent's money. Otherwise, if there is a link from i to j, the node j gains μ_i fraction of i-th agent's money. In the Monte Carlo simulations, one assigns a random amount of money to agents to start with, such that the average money $M/N = 1$. A pair of agents (nodes) are chosen at random, and, depending on the directionality of the link between them (the sign of a_{ij}), the relevant rule from the above is chosen. This is repeated until a steady state is reached and the money distribution does not change in time. The distribution of money $P(m)$ is obtained by averaging over several ensembles (different random initial distribution of money).

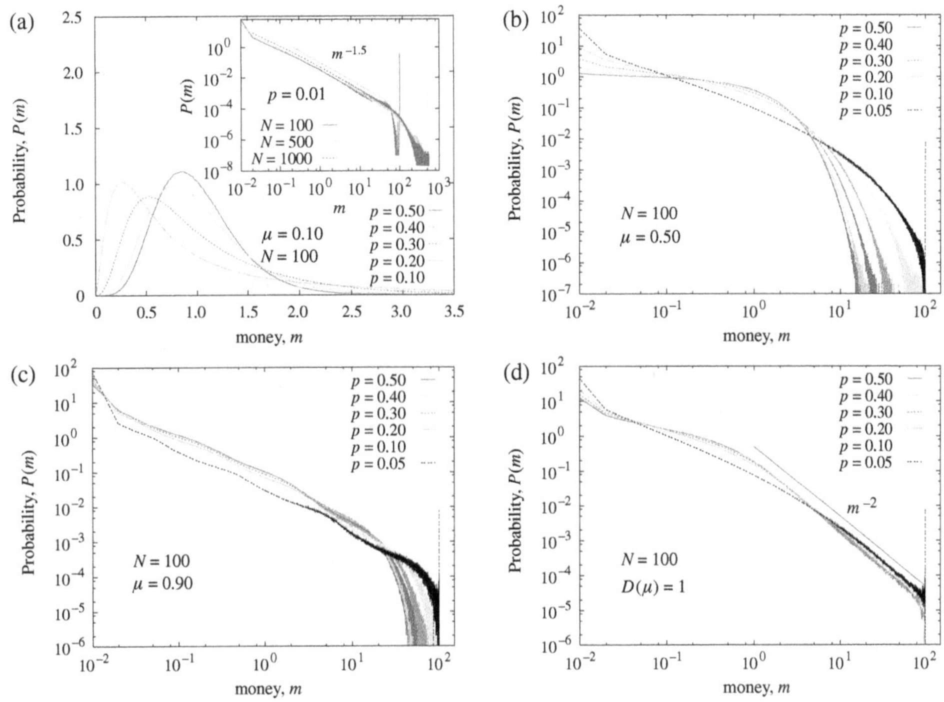

Figure 4.38 The steady-state distribution $P(m)$ of money m for directed networks characterized by different values of p. (a) For $\mu = 0.1$, the inset shows $P(m)$ for $p = 0.01$ for $N = 100, 500, 1000$, and the power law $m^{-1.5}$ is also indicated. (b) For $\mu = 0.5$ and (c) for $\mu = 0.9$. (d) The plots for uniform, random distributed μ, $D(\mu) = 1$, and also a guide to the power law m^{-2}. The data are obtained by numerical simulation, for a system of $N = 100$ nodes, averaged over 10^3 realizations in the steady state and over 10^4 initial configurations. The average money M/N is 1. Reproduced from Chatterjee (2009).

This model is different from the CC and CCM models, but one can relate the transfer fraction μ analogous to λ in the CC and CCM models.

4.3.1.1 Model with uniform μ

We first discuss the case of homogeneous agents, i.e. when all agents i have $\mu_i = \mu$. The $\mu = 0$ limit is trivial, as the system does not have any dynamics. Figure 4.38a shows the steady-state distribution $P(m)$ of money m for $\mu = 0.1$ for different values of network disorder p. In general, the distribution of money has a most probable value, which shifts monotonically from about 0.85 for $p = 0.5$ to 0 as $p \to 0$. $P(m)$ has an exponential tail, but a power law region develops as $p \to 0$ below the exponential cut-off, which fits approximately to $m^{-1.5}$. At $p \to 0$, the condensation of wealth at the node(s) with strong disorder ($\rho \to 1$) is apparent from the single, isolated datum point at the $m_{max} = M$ end (see inset of Fig. 4.38a, for $p = 0.01$). There is a strong finite size effect involved in this behaviour. To emphasize this, we plot $P(m)$ for $p = 0.01$ for $N = 100, 500$ and 1000 (inset of Fig. 4.38a). While $N = 100$ and $N = 500$ show the isolated datum point, it is absent for $N = 1000$. This also indicates that this behaviour is absent for infinite systems for $p \to 0$. For larger values of p, the distribution resembles gamma distributions, as in the CC model. At $\mu = 0.5$, the most-probable value of $P(m)$ is always at 0 (Fig. 4.38b). For weak disorder ($p = 0.5$), $P(m)$ is exponential, but it shows a wider distribution as one goes to higher disorder ($p \to 0$). The condensation of wealth at node(s) with high value of ρ ($\rho \to 1$) is again apparent from the single, isolated datum point at the $m_{max} = M$ end (Fig. 4.38b, for $p = 0.05$). For $\mu = 0.9$, $P(m)$ is always decaying, with a wide distribution up to $m_{max} = M$ (Fig. 4.38c). As in previous plots, the condensation of wealth at node(s) with high value of ρ is visible: see plot for $p = 0.05$ in Fig. 4.38c. A common feature for the curves for all values of p is that $P(m)$ exhibits log-periodic oscillations, while resembling roughly a power law decay. Another important feature is that $P(m) \to N$ for $m \to 0$, which indicates that money is distributed in a very small fraction of nodes, while most nodes have almost no money at a given instance.

For a particular value of the network disorder p, the wealth distribution $P(m)$ becomes more and more 'fat tailed' as μ is increased. This is in contrast to what is observed in the CC model (Chakraborti and Chakrabarti 2000) where $P(m)$ organizes to a narrower distribution as λ increases.

4.3.1.2 *Model with distributed* μ

We now consider the case in which each agent i has a different value of μ_i, which does not change in time. This is a case of heterogeneous agents where the heterogeneity can be viewed as a 'quenched disorder'. We consider a random uniform distribution of μ, i.e. $D(\mu) = 1$ in $0 < \mu < 1$. This is the case analogous to the CCM model (Chatterjee *et al.* 2004). Figure 4.38d shows the plots of the money distribution $P(m)$ for different values of network disorder p. All curves have a power law tail (Chatterjee 2009), resembling a m^{-2} variation. However, the effects of the topology of the underlying network are visible: for strong disorder in topology $p = 0.05$, condensation of wealth at node(s) with a high value of ρ ($\rho \to 1$) is also apparent from the single, isolated datum point at the $m_{max} = M$ end. Further investigations also indicate that the power law exponent is similarly related to the distribution of 'transfer fraction' μ, as one observes in the CCM model (Chatterjee *et al.* 2004; Mohanty 2006), i.e. $P(m) \sim m^{-2}$ for most distributions, while one can obtain $P(m) \sim m^{-(2+\delta)}$ if $D(\mu) \propto (1 - \mu)^{\delta}$.

For a particular value of the network disorder p, the wealth distribution $P(m)$ becomes more and more 'fat tailed' as μ is increased. Again, this is in contrast to

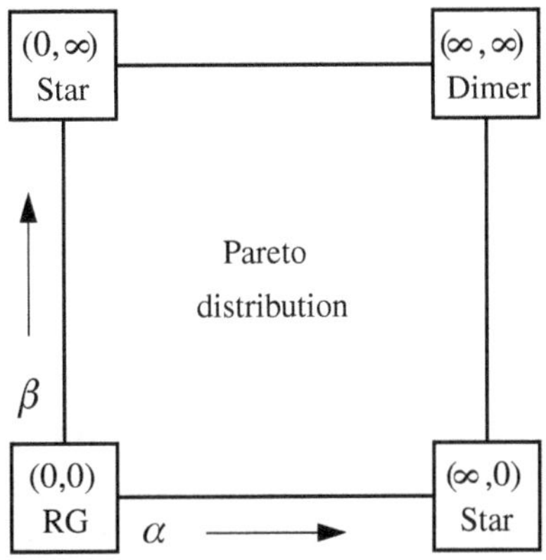

Figure 4.39 The phase diagram in the (α, β) plane. The origin corresponds to the CCM model, while, at corners $(0, \infty)$ and $(\infty, 0)$, the richest trader participates in every transaction, hence the network is star-like. At the (∞, ∞) corner only the richest and the second richest traders trade, and the network essentially is a dimer.

what is observed in the CC model (Chakraborti and Chakrabarti 2000), in which $P(m)$ organizes to a narrower distribution as λ increases.

4.3.2 Preferential transactions and weighted trade network

Very recently, Chakraborty and Manna (2010) proposed models for preferential transactions. The main idea being the fact that rich traders invest much more in trade and hence take part more frequently in the trading process. In the model, a pair of traders i and j are selected for trading with probabilities directly proportional to $m_i(t)^\alpha$ and $m_j(t)^\beta$, respectively. The trading rules are the same as in Eqs. (4.13) and (4.14). The $\alpha = \beta = 0$ case corresponds to the CCM model, the trading topology is a random graph. But when they are non-zero positive, rich traders have higher probability of getting selected in a trading process. However, in this case it takes a long time for all of the traders to take part in the exchange process.

When $\alpha = \beta$ and finite, the resultant wealth distribution exhibits a power law tail with $\nu = 1.00$ apart from slight variations. This shows that the wealth distribution is robust with respect to the parameter values in the region, and the non-zero values of α and β only control the frequency at which different traders take part in the trading process. When either of α or β is infinity and the other is zero, the richest trader is always selected while all others are selected with uniform probability. This corresponds to a star-like topology in the trade graph. It is observed that Pareto law still holds good. Now, when both (α, β) assume large values, the situation looks very different. In its limiting case, (∞, ∞), only the richest and the next richest trader take part in the trading process, i.e. the trade graph is merely a dimer. Thus, qualitatively, one can summarize the results in a schematic diagram (Fig. 4.39).

4.3.2.1 The trade network

One can associate a network in this trading system: each trader being a node and a link appears between a pair when they trade for the first time. No further link is added between a pair even if they trade again. As more and more traders take part in the trading process, the number of links grows in the system. When $\alpha = \beta = 0$ this growth is exactly the same as a random graph, but it looks much different for $\alpha > 0$, $\beta > 0$, because rich nodes preferentially enter into trading and get linked more often than the poor ones. The degree k_i of a node i is the number of distinct traders with whom it traded.

The dynamics is found to have two distinct time scales: T_1, at which the network is a single component connected graph, and T_2, at which the graph is an N-clique (each node is connected to all others). This study reports the growth of the giant component $\langle s_M(\rho, N)\rangle$, which is the order parameter of this percolation problem with respect to the link density $\rho = n/[N(N-1)]$ in the network. A finite size scaling collapses the order parameter for different sizes: scaling ρ axis by a factor of N^{θ}. The critical density of percolation transition $\rho_c(N)$ is defined to be that value of ρ for which $\langle s_M(\rho, N)\rangle = 1/2$. The paper reports that $\rho_c(N)$ varies with $N^{-\theta}$. In general, the exponent $\theta(\alpha)$ depends on α: for $\alpha \leq 1/2$, $\theta(\alpha) = 1$, while for $\alpha > 1/2$, $\theta(\alpha)$ decreases. For Erdős–Renyi random graphs, $\theta = 1$, and hence it indicates that this trade network is different from random graphs for $\alpha > 1/2$.

4.3.2.2 Degree distribution

The degree distribution shows interesting observations. This paper (Chakraborty and Manna 2010) studies the average degree distribution $P(k, N)$ as a function of k at different system sizes N for different values of α, β. It is observed that almost the entire degree distribution obeys the usual finite-size scaling analysis and confirms the validity of the following scaling form:

$$P(k, N) \propto N^{-\eta_k(\alpha)} \mathcal{G}[k/N^{\zeta_k(\alpha)}], \tag{4.36}$$

where the scaling function $\mathcal{G}(y)$ has its usual forms $\mathcal{G}(y) \sim y^{-\gamma(\alpha)}$ as $y \to 0$ and $\mathcal{G}(y)$ approaches 0 very fast for $y \gg 1$. This is satisfied only when $\gamma_k(\alpha) = \zeta_k(\alpha)/\eta_k(\alpha)$, and exponents $\eta_k(\alpha)$ and $\zeta_k(\alpha)$ fully characterize the scaling of $P(k, N)$. $\gamma_k(\alpha)$ is observed to decrease with α.

4.3.2.3 The weighted network

Between an arbitrary pair of traders, a large number of bipartite trading takes place within a certain time T. The sum of the amounts δ_{ij} invested in all trades between traders I and j within time T is defined as the total volume of trade $w_{ij} = \sum_T \delta_{ij}$. Here w_{ij} is known as the weight of the link (i, j). The probability distribution

$P(w, N)$ of the link weights is calculated when the average degree $\langle k \rangle$ reaches a specific preassigned value. When the trade networks is an N-clique graph, i.e. when each trader has traded with all other traders at least once, each node has same degree, i.e. $P(k) = \delta(k - (N - 1))$ and $\langle k \rangle = N - 1$. The distribution has a very long tail, and $P(w, N) \propto w^{-\gamma_w}$ with $\gamma_w \simeq 2.52$.

The strength of a node $s_i = \sum_j w_{ij}$, where j are all neighbours k_i of i, is a measure of the total volume of trade handled by the i-th node. Nodal strengths vary widely over different nodes. The strength distribution also follows a power law decay $P(s, N) \sim s^{-\gamma_s}$ for $N \to \infty$; similarly as in Eq. (4.36).

Often, weighted networks have non-linear strength degree relations indicating the presence of non-trivial correlations, as in the airport networks and the international trade network. For a network where the link weights are randomly distributed, the $\langle s(k) \rangle$ grows linearly with k. However, a non-linear growth like $\langle s(k) \rangle \sim k^{\phi}$, with $\phi > 1$, exhibits the presence of non-trivial correlations. For this case, $\phi(\alpha)$ increases with α. The paper also reports the variation of the mean wealth of a trader with its degree, and gets: $\langle x(k) \rangle \sim k^{\mu(\alpha)}$, where $\mu(\alpha)$ decreases with α.

4.4 Models with debt

What happens when debt is permitted? This is a very important question when it comes to individual economic activity, and debt may be simply considered as negative money. For instance, a bank can be considered as a huge reservoir of money. When an agent borrows money from it, its *cash* balance increases, at the cost of a debt obligation to the bank, which is actually negative money. Thus, a general conservation law is still maintained for the total money, which is the algebraic sum of the cash M and the debt D, i.e. $M - D = M_b$ (Yakovenko and Barkley Rosser 2009). Thus, relaxing the condition of negative money allows for a different *ground state* for the agents, other than $m_i = 0$. A detailed discussion of the above and book-keeping accounting under an econophysics framework can be found in Braun (2001) and Fischer and Braun (2003a,b).

Now, if one considers the DY model (Drăgulescu and Yakovenko 2000) and relaxes the non-negativity of money by allowing agents to go for debt, any agent who loses an amount Δm at an instant when its instantaneous money m_i is less than Δm goes into the state of negative money or debt. The probability distribution of money $P(m)$ never stabilizes, and the system never reaches a steady state – $P(m)$ keeps spreading to $m = +\infty$ and $m = -\infty$, but respecting the conservation of the average money per agent. What does this signify in the economics sense? The conservation tells us that some agents become very rich at the expense of some agents going into greater and greater debt. Any economic system with unlimited debt is not stable, as is learnt from the recent financial crisis. It is

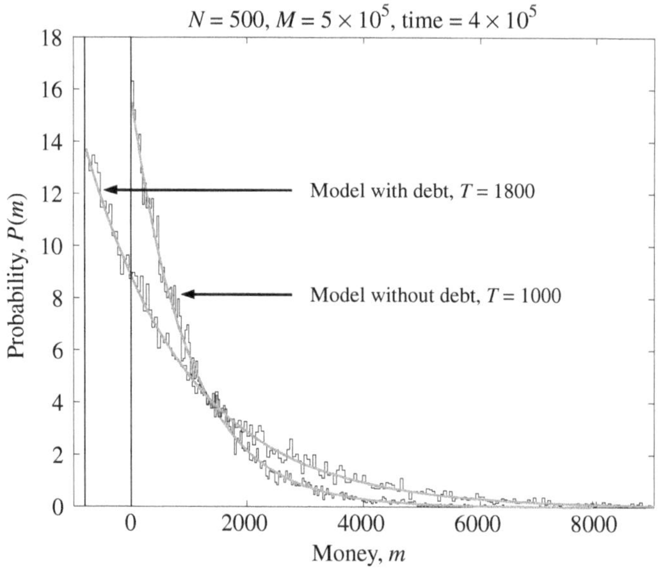

Figure 4.40 Stationary distributions of money with and without debt. The debt cut-off is $m_d = 800$. The solid lines fit to Boltzmann–Gibbs laws with 'money temperature' $T_m = 1800$ and 1000. Reproduced from Drăgulescu and Yakovenko (2000).

commonly believed that the crisis was a result of accumulation of huge amounts of debt.

What happens if the boundary conditions for debt are slightly modified? For example, if there is a limitation on the amount of debt any agent can incur, i.e. instead of the original $m_i \geq 0$, we have a $m_i \geq -m_d$ for all agents i (Drăgulescu and Yakovenko 2000). The resultant stationary distribution of money is still exponential, with respect to the new boundary $m = -m_d$, while the money temperature is now $T_d = m_d + M_b/N$ (Fig. 4.40). This allows each agent to have m_d amount of money at the cost of this debt feature. Another study considers a bit more realistic boundary condition, where, instead of putting a constraint on the individual debt, the constraint is put on the collective debt of the agents in the system (Xi *et al.* 2005). Usually there exists a quantity called the *required reserve ratio* R, which is a fraction of money deposited into bank accounts that a bank sets aside, so that the remaining $1 - R$ fraction can be given away as loans. If initially there is money M_b in the system, then repeated loans and borrowing can increase the money available for loans to $M = M_b/R$. This extra money comes from the increase of total debt in the system. It is easily seen that that maximal total debt is $D = M_b/R - M$. When the debt is maximal, M_b/R positive and $M_b(1 - R)/R$ negative money circulates within the agents, which means that the distributions of positive and negative

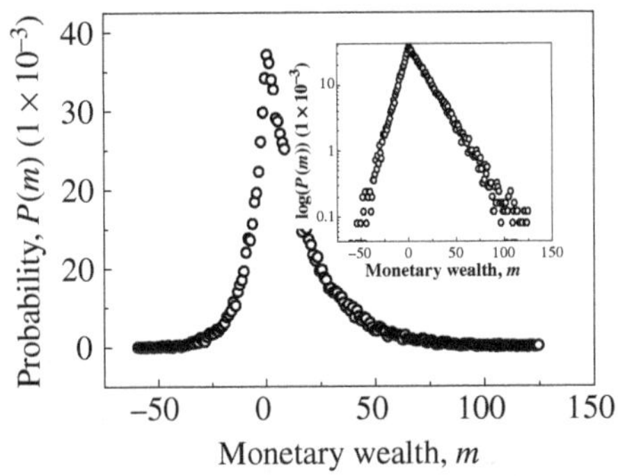

Figure 4.41 Stationary distributions of money for $R = 0.8$, with different 'money temperatures' T_+ and T_-. Reproduced from Xi *et al.* (2005).

monies will be characterized by different 'money temperatures' $T_+ = M_b/RN$ and $T_- = M_b(1 - R)/RN$, as confirmed by simulations of Xi *et al.* (2005) (Fig. 4.41). Similar distributions were also noted elsewhere (Fischer and Braun 2003b). The above framework of the reserve ratio hold for individuals and corporations is not under consideration in reality. In addition, there are other different sources in which debt is accumulated, and sometimes in huge proportions. So far, there has been no consideration of interest rates. Drăgulescu and Yakovenko (2000) considered a model with different fixed interest rates for deposits and loans. Computer simulations revealed that the money distributions are still exponential, but the money temperature varies slowly in time. Another sophisticated model (Keen 1995, 2000) found a regime of debt-induced breakdown, which stops all economic activity due to heavy debt, and requires a 'debt moratorium' or a delay in the payments of debts or obligations, in order to resume any further economic activity. One can go from fixed interest rates, as in the above models, to a scenerio where they are adjusted self-consistently, say, where the interest rates are set according to probabilistic withdrawals of deposits (Cockshott and Cottrell 2008). This study revealed that the money supply first increases up to a certain limit, and then the accumulated debt induces a 'crash' in the economy. These are probably very interesting areas for future studies (Yakovenko and Barkley Rosser 2009).

4.5 Models with tax

Guala (2009) provided a simple framework under which the effect of taxes can be realized in simple wealth exchange models. The basic idea is that there is a process of taxation involved in any trading interaction. The simple model describes a taxation process for all binary interactions, where a fraction of the total money involved is lost as a tax, which could be redistributed according to some prescribed rule. Formally, if f is the *tax* parameter, the amount $f(w_i(t) + w_j(t))$ is given away as taxes, $w_i(t)$ being the wealth of agent i at an instant (trading time step) t. The available wealth $(1 - f)(w_i(t) + w_j(t))$ is split up randomly between the trading agents i and j, similar to Eq. (4.4), as

$$
\begin{aligned}
w_i'(t+1) &= \epsilon_t(1 - f)(w_i(t) + w_j(t)), \\
w_j'(t+1) &= (1 - \epsilon_t)(1 - f)(w_i(t) + w_j(t)),
\end{aligned}
$$

with ϵ_t being random with time t. The *tax* part is redistributed to a subset r, which could be either some poorest fraction of the population or could also be all of them. Guala discusses the simple case when the tax is redistributed within the whole population. For $f = 0$, it is easily seen that the model is nothing but the case of the random sharing of wealth (Drăgulescu and Yakovenko 2000), which gives a purely exponential decay for the steady-state distribution of wealth $P(w)$ with modal value $w_m = 0$. However, for the cases $f \neq 0$, $P(w)$ is asymmetric unimodal with $w_m \neq 0$, but w_m shifts away from 0 for $f = 0$ until it goes to a maximum before decreasing again until it becomes 0 for $f = 1$. The most *egalitarian* distribution is found for an optimal value of the taxation $f \simeq 0.325$.

4.6 Other related models

4.6.1 Inelastic scattering in an open economy

Slanina (2004) introduced a generalized model of kinetic exchanges, but incorporating a non-conservation term. The basic assumption is that the system is open and the interaction can produce an increase of the total wealth of the two interacting agents. The external energy is utilized only through a human activity and the problem is simplified by assuming that the net increase of wealth happens at the very moment of the agents' interaction. Also, the interactions occur pairwise. In each time step t a pair of agents (i, j) is chosen randomly. They interact and exchange wealth (4.1) according to the symmetric rule

$$
\mathcal{M} = \begin{pmatrix} 1 + \epsilon - \beta & \beta \\ \beta & 1 + \epsilon - \beta \end{pmatrix}. \tag{4.37}
$$

All other agents leave their wealth unchanged, $m_k(t + 1) = m_k(t)$ for all k different from both i and j. $\beta \in (0, 1)$ quantifies the wealth exchanged, while $\epsilon > 0$ is the measure of the flow of wealth from the outside. The process is shown in Fig. 4.42. The parameter ϵ takes care of all sources of non-conservation.

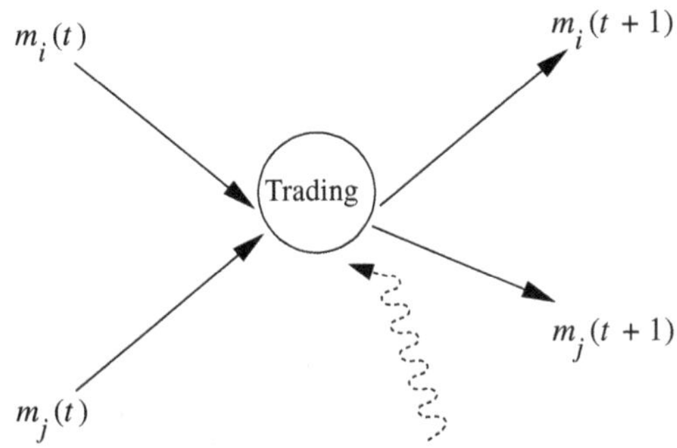

Figure 4.42 Schematic picture of the scattering process, in which the wealth is exchanged and produced.

4.6.2 A threshold-induced phase transition in the kinetic exchange models

In the context of wealth exchange processes, Pianegonda *et al.* (2003) and Iglesias (2010) had considered a model for the economy in which the poorest in the society (atom with least energy in the gas) at any stage takes the initiative to go for a trade (random wealth/energy exchange) with anyone else. Interestingly, in the steady state, one obtained a self-organized poverty line, below which none could be found and above which a standard exponential decay of the distribution (Gibbs) was obtained.

Along the same lines, Ghosh *et al.* (2011) studied a model in which N particles interact among themselves through two-body energy (x) conserving stochastic scatterings with at least one of the particles having energy below a threshold θ (same as the poverty line in the equivalent economic model). The states of particles are characterized by the energy $\{x_i\}, i = 1, 2, \ldots, N$, such that $x_i > 0, \forall i$ and the total energy $E = \sum_i x_i$ is conserved (= N, such that the average energy of the system $\bar{E} = E/N = 1$, without any loss of generality). The evolution of the system is carried out according to the following dynamics:

$$\left.\begin{aligned} x_i^{<\prime} &= \epsilon(x_i^{<} + x_j), \\ x_j^{\prime} &= (1-\epsilon)(x_i^{<} + x_j), \end{aligned}\right\} \tag{4.38}$$

where $x_i^{<} < \theta$ (threshold energy or 'poverty line') and ϵ ($0 \le \epsilon \le 1$) is a *stochastic* variable, changing with time (scattering). The quantity x is conserved during each collision: $x_i^{<\prime} + x_j^{\prime} = x_i^{<} + x_j$. The question of interest is: what is the steady-state distribution $p(x)$ of energy x in such systems?

They simulated a system of N particles (agents). At any time t, a particle i is selected randomly. If its energy is below a prescribed threshold energy θ, it collides with any other random particle j (in the mean-field model), and the two particles will exchange energy according to Eq. (4.38). After each such successful collision, the time is incremented by unity. The dynamics continues for an indefinite period, unless there is no particle left below the threshold energy, in which case the dynamics freezes. After sufficiently large time $t > \tau$, a steady state is reached when the energy distribution $p(x)$ (and also other average quantities) does not change with time. One may start with different initial random configurations, where the states of particles are characterized by the energies $\{x_i\}, i = 1, 2, \ldots, N$, which are drawn randomly from a uniform distribution such that $x_i > 0, \forall i$ and the average energy $\bar{E} = \sum_i x_i/N$ is set to unity. One finds the system to be *ergodic* (the steady-state distribution $p(x)$ is *independent* of the initial conditions $\{x_i\}$), and steady-state averages over all such independent initial conditions is taken to evaluate the quantities of interest.

Ghosh *et al.* (2011) observed that, for finite values of the energy threshold θ, the steady-state energy distribution is no longer the simple Gibbs–Boltzmann distribution. They also found that O ($\equiv \int_0^\theta p(x)dx$), the average number of particles below the threshold energy in the steady state, is zero for θ values below or at a critical threshold energy θ_c, and, for $\theta > \theta_c$, O is non-zero. The steady-state value of O, the average number of particles below the threshold energy θ, is the 'order parameter' of the system. They studied the relaxation dynamics in the system: the relaxation of $O(t)$ to the steady-state value of O ($= O(\theta)$ for $t > \tau(\theta)$, the 'relaxation time'). They found $\tau(\theta)$ grows as θ approaches θ_c, and eventually diverges at θ_c. They studied mainly three cases: (1) the mean-field (or infinite range) case where i and j in Eq. (4.38) can represent any two particles/agents in the system; (2) the one-dimensional case where $j = i \pm 1$ along a chain; and (3) the two-dimensional case in which $j = i \pm \delta$, where δ represents neighbors of i; they considered a two-dimensional square lattice.

In the mean-field (infinite range) model, the order parameter $O \equiv \int_0^\theta p(x)dx$ (Fig. 4.43) shows a 'phase transition' at $\theta_c \simeq 0.607 \pm 0.001$. A power law fit $O \sim (\theta - \theta_c)^\beta$ gives $\beta \simeq 0.97 \pm 0.01$.

They also studied the relaxation behaviour of O. At $\theta = \theta_c$, the $O(t)$ variation fits well with $t^{-\delta}$; $\delta \simeq 0.93 \pm 0.01$. The relaxation time τ was also estimated numerically and found a diverging growth of τ near $\theta = \theta_c$ (Fig. 4.44), indicating 'critical slowing down' at the critical value θ_c. The values of exponent z for the divergence in $\tau \sim |\theta - \theta_c|^{-z}$ were estimated (for both $\theta > \theta_c$ and $\theta < \theta_c$). For the mean-field model, the fitting value for exponent $z \simeq 0.83 \pm 0.01$. They also studied the universality of this behaviour by generalizing the dynamics in Eq. (4.38) to

$$\left.\begin{aligned} x_i^{<\prime} &= \epsilon_1 x_i^< + \epsilon_2 x_j, \\ x_j' &= (1 - \epsilon_1) x_i^< + (1 - \epsilon_2) x_j, \end{aligned}\right\} \tag{4.39}$$

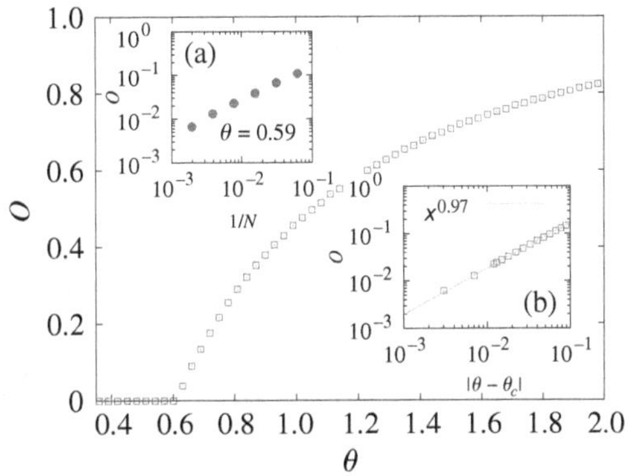

Figure 4.43 Simulation results for the variation of O, the average number of particles below the threshold energy θ in the steady state ($t > \tau$), against threshold energy θ. (Inset)(a) The results at $O \to 0$ for $\theta = 0.59 (< \theta_c)$ as $N \to \infty$; (b) scaling fit $(\theta - \theta_c)^\beta$ with $\beta \simeq 0.97$. Simulations are for the mean-field model with $N = 10^5$. Reproduced from Ghosh *et al.* (2011).

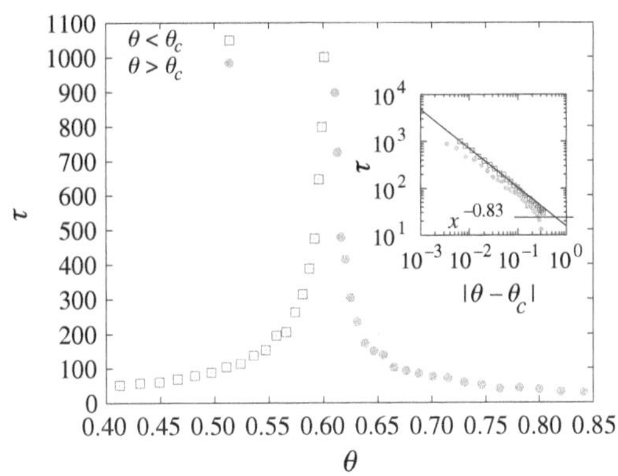

Figure 4.44 Variation of τ versus θ. (Inset) Scaling fit $\tau \sim |\theta - \theta_c|^{-z}$, with exponent $z \simeq 0.83$. Simulations are shown for the mean-field case with $N = 10^5$. Reproduced from Ghosh *et al.* (2011).

where ϵ_1 and ϵ_2 are random stochastic variables within the range [0, 1]. The critical point θ_c shifted to $\theta_c \simeq 0.69$ ($\theta_c \simeq 0.61$ for $\epsilon_1 = \epsilon_2 = \epsilon$). The transition behaviour was seen to be universal near the critical point θ_c, but the critical point depended specifically on the model.

In the one-dimensional version, they found that $\beta \simeq 0.41 \pm 0.02$ and $\theta_c \simeq 0.810 \pm 0.001$, while z turned out to be around 1.9 ± 0.05 and $\delta \simeq 0.19 \pm 0.01$. For the two-dimensional lattice version, $\beta \simeq 0.67 \pm 0.01$ and $\theta_c \simeq 0.675 \pm 0.005$. In addition, they found $z \simeq 1.2 \pm 0.01$ and $\delta \simeq 0.43 \pm 0.02$. All these estimated values of the critical exponents β, z and δ are summarized in Table 4.1.

Table 4.1 *Comparison of critical exponents of the Ghosh* et al. *model with those of the Manna model (Lübeck 2004)*

		Ghosh *et al.*	Manna
β	1D	0.41 ± 0.02	0.382 ± 0.019
	2D	0.67 ± 0.01	0.639 ± 0.009
	MF	0.97 ± 0.01	1
z	1D	1.9 ± 0.05	1.876 ± 0.135
	2D	1.2 ± 0.01	1.22 ± 0.029
	MF	0.83 ± 0.01	1
δ	1D	0.19 ± 0.01	0.141 ± 0.024
	2D	0.43 ± 0.02	0.419 ± 0.015
	MF	0.93 ± 0.01	1

1D, one-dimensional; 2D, two-dimensional; MF, mean field.

In summary, when the energy threshold θ was introduced in the kinetic theory of an ideal gas, the stochastic energy-conserving scatterings between any two particles could take place only when one has energy less than θ. The system showed a 'phase transition' at $\theta = \theta_c$, with exponent values given in Table 4.1.

Chakraborty *et al.* (2011) considered a similar version of the model by Pianegonda *et al.* (2003), which is driven by the idea of extremal dynamics. In their model, the agent with globally minimal wealth w_{min} is selected, and its *neighbour* for bipartite transaction can be selected randomly and uniformly using various rules, under various definitions of the interaction topology in terms of 1d and 2d with periodic boundary conditions, infinite range (mean field/N-clique) and Barabasi–Albert scale-free networks. Initially the agents are given random amounts of wealth w_i $(i = 1, 2, \ldots, N)$ such that the average wealth is unity, $\langle w \rangle = 1$. A pair of agents selected in the above manner pool their wealth and randomly share it. This process continues forever. After some relaxation time, the system reaches a stationary state when the wealth distribution assumes a time-independent form. It is observed that none of the agents have wealth below a certain value for an infinite system (suing finite-size scaling arguments). Thus a 'forbidden' region of wealth is formed and agents, owing to the dynamics, cannot end up with wealth below a certain cut-off which is generated in a self-organized manner. Similarly, the dynamics can be redefined such that the agent with the maximum wealth is chosen for trading, and this produces a cut-off in the wealth owing to a self-organized dynamics such that no agent has wealth above it. The details of the critical behaviour of these models are reported in Chakraborty *et al.* (2011). Although the wealth distributions produced in these models hardly resemble any known form seen in empirical data, the model dynamics gives us a new insight into the possible processes that lead to such distributions.

Very recently, Iglesias and de Almeida (2012) have presented a study of a family of wealth exchange models and shown that, when the exchange rule is not time-reversible and does not allow the poorer agent in the interaction to win more than its own endowments, the system converges to a state of *minimum* entropy, with full condensation of the wealth in a few hands, and the termination of trade. They also calculated the Lorenz curves and the Theil and Gini coefficients. Finally, they proposed that only a rule where the poorer agent is favoured in a very significant way produces a wealth distribution with less inequality and a Gini coefficient lower than 1.

In this chapter we have considered the market exchanges and scattering models, and their numerical studies. The details of the analytical solutions of the kinetic exchange models will be presented in the following chapter.

5

Analytic structure of the kinetic exchange market models

The kinetic exchange models of markets provide a simplified – perhaps oversimplified – picture of the exchange mechanism that takes place in a real market. However, the simple, 'toy model' paradigm also offers interesting grounds for possible analytic formulation, compared with real markets where the dynamics involve plenty of parameters and involve complex evolution that render them intractable and incomprehensible.

In this chapter we will discuss in detail some of the simple and intuitive frameworks developed to understand the qualitative, and sometimes even the quantitative, aspects of simple kinetic exchange models discussed in the previous chapters. Most of our discussions will include the CC (Chakraborti and Chakrabarti 2000) and the CCM (Chatterjee *et al.* 2004) models, and some of their important variants, which are easy to handle analytically, or in cases where their solutions can be argued intuitively.

5.1 Analytic results for the CC model

The earliest attempt to understand the CC model analytically (Das and Yarlagadda 2003) assumes that, independent of the initial conditions, the system evolves to an equilibrium distribution after a sufficiently long time. Thus, in the steady state, the joint probability that, before interaction, money of i lies between x and $x + dx$ and that of j lies between z and $z + dz$ is $f(x)dxf(z)dz$. Since each interaction conserves total money, let $L = x + z$. Then the joint probability becomes $f(x)dxf(L - x)dL$. The probability that L is distributed to give money of i between y and $y + dy$ is

$$\frac{dy}{(1-\lambda)L} f(x)dxf(L-x)dL, \tag{5.1}$$

with $x\lambda \leq y \leq x\lambda + (1-\lambda)L$. It is evident that $x \leq y/\lambda$ and $x \geq [y - (1 - \lambda)L]/\lambda$. x satisfies the constraint $0 \leq x \leq L$ because the agents cannot have negative money. Thus the upper limit on x is $\min\{L, y/\lambda\}$ (i.e. minimum of L

and y/λ) and the lower limit is $\max\{0, [y - (1-\lambda)L]/\lambda\}$. Now, the total money L has to be greater than y so that the agents have non-negative money. Thus we have the following distribution function for the money of i to lie between y and $y + dy$

$$f(y)dy = dy \int_y^\infty \frac{dL}{(1-\lambda)L} \int_{\max[0,\{y-(1-\lambda)L\}/\lambda]}^{\min[L,y/\lambda]} dx f(x) f(L-x). \qquad (5.2)$$

It is interesting to observe that, when $\lambda = 0$, the lower and upper limits of the x integration become 0 and L, respectively, giving

$$f(y)dy = dy \int_y^\infty \frac{dL}{L} \int_0^L dx f(x) f(L-x). \qquad (5.3)$$

For the zero savings case, note that the double derivative with respect to y of the above Eq. (5.3) yields

$$f'(y) + y f''(y) = -f(y)f(0) - \int_0^y f(x) f'(y-x) dx. \qquad (5.4)$$

For small y one assumes that the function $f(y)$ and its first and second derivatives are well behaved. Then as $y \to 0$, $f'(y) \approx -f(y)f(0)$. Then the solution for small y, after using the constraint $\int_0^\infty f(y)dy = 1$, is given by

$$f(y) \approx f(0) \exp[-y f(0)]. \qquad (5.5)$$

One can see that the above function (Eq. (5.5)) is also a solution of the parent Eq. (5.3).

This work (Das and Yarlagadda 2003) also draws an important conclusion from Eq. (5.2): assuming $f(y) \to 0$ as $y \to \infty$ and that $f(y)$ is well behaved as $y \to 0$, it is clear that $\lim_{y\to 0} f(y) \to 0$. Physically this is plausible because if everyone saves, then, owing to interactions, a person with zero money will tend to finite money faster than returning to zero money. A person with zero money after a single interaction has unit probability of having non-zero money. And, once a person has non-zero money it takes an infinite number of interactions to lose and end up with zero money. Thus the $\lambda = 0$ case and $\lambda > 0$ case belong to different 'universality classes'. Moreover $\lambda = 1$ is different because it is a static situation. The authors also argued that for $0 < \lambda < 1$ as $y \to \infty$ the function decays exponentially. They solved the integral equation given by Eq. (5.2) for the non-trivial case of $f(y) \neq \delta(y)$, and obtained the series of curves very similar to Fig. 4.2.

5.1.1 Does the CC model produce gamma distribution?

In a very interesting paper by Matthes and Toscani (2008b), an argument was put forward whether the series of curves for the density of money distribution for the CC model resemble gamma distributions. It all started with an observation that the equilibrium distribution for a given λ $(0 \leq \lambda < 1)$ fits extremely well to the data from numerical simulations. The function is conveniently expressed in terms of the parameter

$$n(\lambda) = 1 + \frac{3\lambda}{1-\lambda}. \tag{5.6}$$

This particular form of $n(\lambda)$ was also suggested by a mechanical analogy (Chakraborti and Patriarca 2008), between the closed economy model with N agents and the dynamics of a gas of N interacting particles. Then the money distributions, for arbitrary values of λ, were well fitted by the function

$$\left.\begin{aligned} f_n(x) &= a_n x^{n-1} \exp\{-nx/\langle x\rangle\}, \\ a_n &= \frac{1}{\Gamma(n)} \left(\frac{n}{\langle x\rangle}\right)^n, \end{aligned}\right\} \tag{5.7}$$

where n is defined in Eq. (5.6) and the prefactor a_n, where $\Gamma(n)$ is the gamma function, is fixed by the normalization $\int_0^\infty dx f_n(x) = 1$. Using a rescaled variable

$$\xi = nx/\langle x\rangle, \tag{5.8}$$

the probability distribution (5.7) can be rewritten as

$$\frac{\langle x\rangle}{n} f_n(x) = \frac{1}{\Gamma(n)} \xi^{n-1} \exp\{-\xi\} \equiv \gamma_n(\xi), \tag{5.9}$$

where $\gamma_n(\xi)$ is the standard gamma distribution (Ross 1970). The cumulative distribution for $\gamma_n(\xi)$ is the incomplete gamma function $\Gamma(\xi, n) = \int_\xi^\infty d\xi' \, \gamma_n(\xi')$,

$$\gamma_n(\xi) \equiv -\frac{d}{d\xi} \frac{\Gamma(\xi, n)}{\Gamma(n)}. \tag{5.10}$$

The Gibbs distribution

$$f_1(x) = \frac{1}{\langle x\rangle} \exp\left\{-\frac{x}{\langle x\rangle}\right\} \tag{5.11}$$

is a special case for $n = 1$. The term $\langle x\rangle/n$, on the left-hand side of Eq. (5.9), is just the scaling factor appearing when the change of variable, from ξ to x, is made in the last equation above in order to obtain the distribution $f_n(x)$ for the variable x.

The distribution defined by Eqs. (5.7) has the following characteristics, compared with the Gibbs distribution:

- The power x^{n-1} and the factor n in the exponential, which qualitatively change the distribution shape; x_m, the mode, which is different from zero, goes well with the theoretical prediction that the mode $x_m = 3\lambda/\{1+2\lambda\}$ is obtainable from Eq. (5.7).
- The presence of the factor n is relevant for a mechanical analogy.
- As $\lambda \to 1$ (i.e. $n \to \infty$), the distribution $f_n(x)$ tends to a Dirac δ-function, peaked around the average value $\langle x \rangle$.

Moreover, the curves $f_n(x)$'s, for different values of λ going from zero to unity, are extremely close to those obtained by numerical simulations.

A more rigorous derivation of the asymptotic distribution for $\lambda \to 1$ can be made by studying the characteristic function $\phi(q)$. The gamma distribution $\gamma_1(\xi)$ for the dimensionless variable ξ has a characteristic function $\phi_1(q) = \int_0^{+\infty} d\xi \, \exp(iq\xi)\gamma_1(\xi) = 1/(1-iq)$. The characteristic function of the Gibbs distribution (5.11) $f_1(x)$ is obtained by rescaling q by the constant factor $x/\xi = \langle x \rangle$,

$$\phi_1(q) = \{1 - iq\langle x \rangle\}^{-1}. \tag{5.12}$$

The characteristic function of the generic gamma distribution $\gamma_n(\xi)$ is simply given by the n-th power of $\phi_1(q)$ (Ross 1970), $\phi_n(q) = 1/(1-iq)^n$. Analogously, the corresponding characteristic function of $f_n(x)$ (Eq. (5.7)) is obtained by scaling q by $x/\xi = \langle x \rangle/n$,

$$\phi_n(q) = \{1 - iq\langle x \rangle/n\}^{-n}. \tag{5.13}$$

Thus, in the limit $n \to \infty$ ($\lambda \to 1$), one gets

$$\phi_n(q) \to \exp\{iq\langle x \rangle\}. \tag{5.14}$$

The corresponding distribution is obtained by transforming back the characteristic function, i.e.

$$f_n(x) = (2\pi)^{-1} \int_{-\infty}^{+\infty} dq \, \exp(-iqx)\phi_n(q) \to \delta(x - \langle x \rangle). \tag{5.15}$$

This limit shows that a large saving criterion leads to a final state in which economic agents tend to have the same amount of money and, in the limit of $\lambda \to 1$, they all have the same amount $\langle x \rangle$.

5.1.1.1 *Fixed-point distribution*

Lallouache *et al.* (2010b) obtained some interesting analytical results, in the thermodynamic limit ($N \to \infty$). When the number of agents is very large, a particular agent interacts with another agent very rarely (because agents are chosen randomly with the same probability). Thus, the agents can be considered independent; so, at equilibrium, while the agents' wealth follows the dynamics of Eqs. (4.7)–(4.8), the global distribution does not change, and one can write:

$$X \stackrel{\mathrm{d}}{=} \lambda X_1 + \epsilon(1-\lambda)(X_1 + X_2), \tag{5.16}$$

where $\stackrel{\mathrm{d}}{=}$ means identity in distribution and one assumes that the random variables X_1, X_2 and X have the same probability law, while the variables X_1, X_2 and ϵ are stochastically independent. It is difficult to find the distribution of X; however, one can compute the moments of f. Indeed with Eq. (5.16), one can write immediately

$$\forall m \in \mathbb{N}, \left\langle X^m \right\rangle = \left\langle (\lambda X_1 + \varepsilon(1-\lambda)(X_1 + X_2))^m \right\rangle, \tag{5.17}$$

and by developing Eq. (5.17) one can find the recursive relation

$$\left\langle X^m \right\rangle = \sum_{k=0}^{m} \binom{m}{k} \frac{\lambda^{m-k}(1-\lambda)^k}{k+1} \sum_{p=0}^{k} \binom{k}{p} \left\langle X^{m-p} \right\rangle \left\langle X^p \right\rangle. \tag{5.18}$$

Using Eq. (5.18) with initial conditions $\left\langle X^0 \right\rangle = 1$ (normalization) and $\left\langle X^1 \right\rangle = 1$ (without loss of generality), one obtains

$$\left\langle X^2 \right\rangle = \frac{\lambda + 2}{1 + 2\lambda}, \tag{5.19}$$

$$\left\langle X^3 \right\rangle = \frac{3(\lambda + 2)}{(1 + 2\lambda)^2}, \tag{5.20}$$

$$\left\langle X^4 \right\rangle = \frac{72 + 12\lambda - 2\lambda^2 + 9\lambda^3 - \lambda^5}{(1 + 2\lambda)^2(3 + 6\lambda - \lambda^2 + 2\lambda^3)}. \tag{5.21}$$

As mentioned earlier, Patriarca *et al.* (2004) conjectured that the steady-state distribution for the CC model is the gamma distribution. Now one can compare the moments calculated from Eq. (5.18) with the moments of conjecture Eq. (5.9). Setting $\langle x \rangle = 1$ in Eq. (5.9), one can show

$$\left\langle x^k \right\rangle = \frac{(n+k-1)(n+k-2)\cdots(n+1)}{n^{k-1}}. \tag{5.22}$$

Writing (5.22) for $k = 2, 3, 4$ and choosing n as in Eq. (5.6) we find

$$\left\langle x^2 \right\rangle = \frac{n+1}{n} = \frac{\lambda + 2}{1 + 2\lambda}, \tag{5.23}$$

$$\left\langle x^3 \right\rangle = \frac{(n+2)(n+1)}{n^2} = \frac{3(\lambda + 2)}{(1 + 2\lambda)^2}, \tag{5.24}$$

$$\left\langle x^4 \right\rangle = \frac{(n+3)(n+2)(n+1)}{n^3} = \frac{3(\lambda + 2)(4 - \lambda)}{(1 + 2\lambda)^3}. \tag{5.25}$$

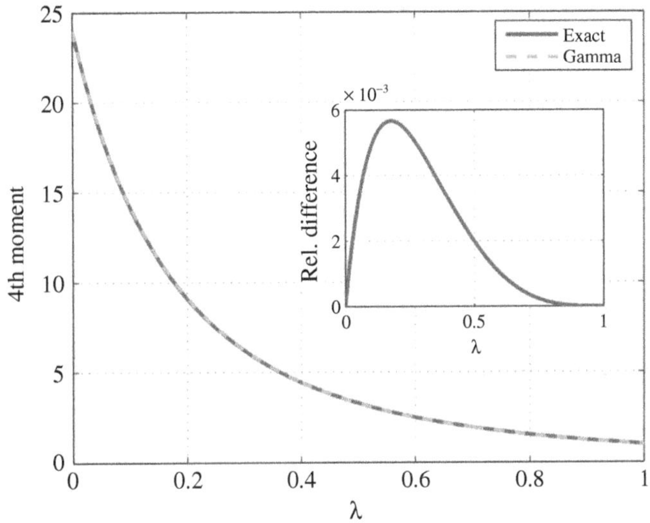

Figure 5.1 Exact fourth moment (Eq. (5.21)) and gamma distribution fourth moment (Eq. (5.25)) against λ. The inset shows the relative difference between the exact fourth moment (Eq. (5.21)) and the gamma distribution fourth moment (Eq. (5.25)) against λ. Reproduced from Lallouache *et al.* (2010b).

One immediately finds that the fourth moments (Eq. (5.21) and Eq. (5.25)) are different, so the conjecture that the gamma distribution is an equilibrium solution of this model is wrong! Nevertheless the first three moments coincide exactly, which shows that the gamma distribution is strangely a *very good approximation*. Moreover the deviation in the fourth moment is very small (see Fig. 5.1, which shows that the two curves can hardly be distinguished by the naked eye). Finding a function that would coincide to higher moments is still an open challenge. These results were found to be consistent with those found by Repetowicz *et al.* (2005), which are presented below.

5.1.1.2 Laplace transform analysis

Lallouache *et al.* (2010b) confirmed the previous result with a different approach based on the Boltzmann equation and along the work of the Toscani group (Matthes and Toscani 2008b; Bassetti and Toscani 2010). Given a fixed number of N agents in a system, which are allowed to trade, the interaction rules describe a stochastic process of the vector variable $(x_1(\tau), \ldots, x_N(\tau))$ in discrete time τ. Processes of this type have been thoroughly studied, e.g. in the context of the kinetic theory of ideal gases. Indeed, if the variables x_i are interpreted as energies corresponding to the i-th particle, one can map the process to the mean-field limit of the Maxwell model of particles undergoing random elastic collisions. The full information about the process in time τ is contained in the N-particle joint probability distribution $P_N(\tau, x_1, x_2, \ldots, x_N)$. However, one can write a kinetic equation for the one-marginal distribution function

$$P_1(\tau, x) = \int P_N(\tau, x, x_2, \ldots, x_N) dx_2 \cdots dx_N,$$

involving only one- and two-particle distribution functions

$$\begin{aligned} &P_1(\tau+1, x) - P_1(\tau, x) \\ &= \left\langle \frac{1}{N} \left[\int P_2(\tau, x_i, x_j) \Big(\delta(x - \lambda x - (1-\lambda)\epsilon(x_i + x_j)) \right.\right. \\ &\left.\left. \quad + \delta(x - \lambda x - (1-\lambda)(1-\epsilon)(x_i + x_j)) \Big) dx_i dx_j - 2P_1(\tau, x) \right] \right\rangle, \end{aligned}$$

which may be continued to give eventually an infinite hierarchy of equations of BBGKY (Born–Bogoliubov–Green–Kirkwood–Yvon) type (Plischke and Bergersen 2006). In the thermodynamic limit, the agents become independent, as explained earlier. Thus, one can write

$$P_2(\tau, x_i, x_j) = P_1(\tau, x_i) P_1(\tau, x_j),$$

which implies a closure of the hierarchy at the lowest level. Therefore, the one-particle distribution function bears all information. Rescaling the time as $t = \frac{2\tau}{N}$ in the thermodynamic limit $N \to \infty$, one obtains for the one-particle distribution function $f(t, x)$ the Boltzmann-type kinetic equation

$$\begin{aligned} \frac{\partial f(t,x)}{\partial t} &= \frac{1}{2} \left\langle \int f(t, x_i) f(t, x_j) \Big(\delta(x - \lambda x - (1-\lambda)\epsilon(x_i + x_j)) \right. \\ &\left. \quad + \delta(x - \lambda x - (1-\lambda)(1-\epsilon)(x_i + x_j)) \Big) dx_i \, dx_j \right\rangle - f(t, x). \end{aligned} \quad (5.26)$$

This equation can be written (Matthes and Toscani 2008b; Bassetti and Toscani 2010) as

$$\frac{\partial f(t,x)}{\partial t} = Q(f, f),$$

where Q is a *collision operator*. A collision operator is bilinear and satisfies, for all smooth functions $\phi(x)$,

$$\begin{aligned} &\int_0^\infty Q(f, f) \phi(x) dx \\ &= \frac{1}{2} \left\langle \int_0^\infty \int_0^\infty (\phi(x_i') + \phi(x_j') - \phi(x_i) - \phi(x_j)) f(x_i) f(x_j) dx_i dx_j \right\rangle, \end{aligned} \quad (5.27)$$

where x_i' and x_j' are the post-trade wealth. With this property, Eq. (5.26) can be written in the weak form, for all smooth functions $\phi(x)$,

$$\frac{d}{dt}\int_0^\infty f(t,x)\phi(x)dx$$
$$= \frac{1}{2}\left\langle \int_0^\infty \int_0^\infty (\phi(x_i') + \phi(x_j') - \phi(x_i) - \phi(x_j)) f(x_i) f(x_j) dx_i dx_j \right\rangle. \quad (5.28)$$

It is very useful because the choice $\phi(x) = e^{-sx}$ gives (after some calculations) the Boltzmann equation for the Laplace transform $\hat{f}$ of f

$$\frac{\partial \hat{f}(t,s)}{\partial t} + \hat{f}(t,s) = \frac{1}{2}\langle \hat{f}(t,(\lambda + (1-\lambda)\epsilon)s)\hat{f}(t,(1-\lambda)\epsilon s)$$
$$+ \hat{f}(t,(1-\lambda)(1-\epsilon)s)\hat{f}(t,1-(1-\lambda)\epsilon s)\rangle. \quad (5.29)$$

For the steady state, and if ϵ is drawn randomly from a uniform distribution, the previous equation reduces to

$$s\hat{f}(s) = \frac{1}{1-\lambda}\int_0^{(1-\lambda)s} \hat{f}(\lambda s + y)\hat{f}(y)dy, \quad (5.30)$$

which coincides with the results of Repetowicz *et al.* (2005). The Taylor expansion of $\hat{f}(s)$ can be derived by substituting the expansion $\hat{f}(s) = \sum_{p=0}^{\infty}(-1)^p m_p s^p$ in Eq. (5.30). Since $\hat{f}(-s)$ is the moment-generating function, we have $\langle x^k \rangle = m_k \cdot k!$. With this method Repetowicz *et al.* (2005) obtained the recursive formula

$$m_p = \sum_{q=0}^{p} m_q m_{p-q} \tilde{C}_q^{(p)}(\lambda) \quad (5.31)$$

with

$$\left.\begin{aligned} \tilde{C}_q^{(p)}(\lambda) &= \frac{\int_0^{(1-\lambda)} (\lambda + \eta)^q \, \eta^{p-q} d\eta}{1-\lambda}, \\ \tilde{C}_{q+1}^{(p)} &= \frac{(1-\lambda)^{p-q-1} - (q+1)\tilde{C}_q^{(p)}}{p-q}, \\ \tilde{C}_0^{(p)} &= \frac{(1-\lambda)^p}{p+1}. \end{aligned}\right\} \quad (5.32)$$

With the above formula one can obtain the first four moments and they match those found in the previous Eqs. (5.19)–(5.21), which confirms that the gamma distribution is not the stationary distribution.

5.1.2 Upper bound form at low wealth range

Lallouache *et al.* (2010b) also presented a calculation for the upper bound at low wealth range. From Eq. (5.16)

$$X \stackrel{d}{=} \lambda X_i + \epsilon(1-\lambda)(X_i + X_j),$$

one has for all $x \geq 0$

$$\mathbb{P}[X \leq x] = \mathbb{P}[\lambda X_i + \epsilon(1-\lambda)(X_i + X_j) \leq x], \tag{5.33}$$

where $\mathbb{P}[.]$ denotes the probability of the event inside the brackets. Again, one considers that the agents are independent, which is true when $N \to \infty$. Then

$$\begin{aligned}\int_0^x dx f(x) = \int_0^\infty dx_i f(x_i) \int_0^\infty dx_j f(x_j)\\ \times \int_0^1 d\epsilon \Theta[x - \lambda x_i + \epsilon(1-\lambda)(x_i + x_j)],\end{aligned} \tag{5.34}$$

where Θ is the Heaviside step function. Taking the derivative with respect to x in both sides, one has

$$f(x) = \int_0^\infty dx_i f(x_i) \int_0^\infty dx_j f(x_j) \int_0^1 d\epsilon \delta[x - \lambda x_i - \epsilon(1-\lambda)(x_i + x_j)]. \tag{5.35}$$

This equation is an integral equation for $f(x)$, which has no solution in closed form. However, one can simplify the equation, by doing the integral over ϵ. The δ-function contributes only for the following constraints:

$$0 \leq x_i \leq x/\lambda, \tag{5.36}$$

$$\frac{x - x_i}{1-\lambda} \leq x_j, \tag{5.37}$$

$$0 \leq x_j. \tag{5.38}$$

The range defined by these constraints is shown in Fig. 5.2. In this range, the derivative of the argument of the delta function with respect to ϵ is $(x_i + x_j)(1-\lambda)$. Hence, one gets

$$f(x) = \frac{1}{1-\lambda} \int_0^{x/\lambda} dx_i f(x_i) \int_{\max\left(\frac{x-x_i}{1-\lambda}, 0\right)}^\infty dx_j f(x_j) \frac{1}{x_i + x_j}. \tag{5.39}$$

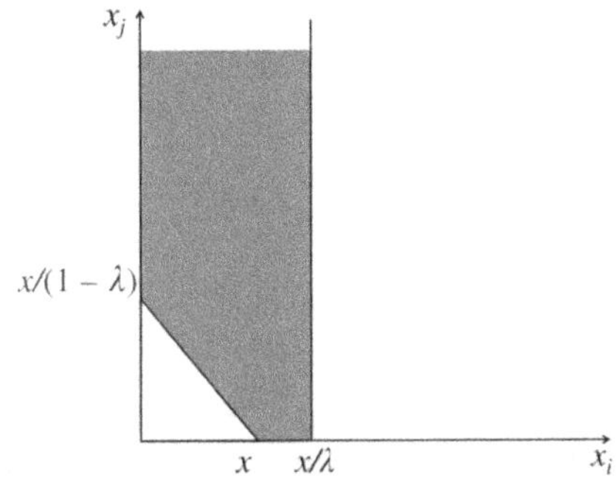

Figure 5.2 Region of integration. Reproduced from Lallouache *et al.* (2010b).

This immediately gives

$$f(x) \leq C \int_0^{x/\lambda} f(x_i) dx_i, \tag{5.40}$$

where

$$C = \frac{1}{1-\lambda} \int_0^\infty dx_j f(x_j) \frac{1}{x_j}. \tag{5.41}$$

One assumes that f decays fast enough near 0, so that the integral in Eq. (5.41) is well defined. Then Eq. (5.40) may be rewritten by rescaling the variable, as

$$f(\lambda x) \leq C \int_0^x dx_i f(x_i). \tag{5.42}$$

If one uses the observation that for $\lambda > 0$ the numerically determined $f(x)$ is a continuous function with a single maximum, say at x_0, then, for all $x \leq x_0$, the integrand (Eq. (5.42)) takes its maximum value at the right extreme point, i.e. when $x_i = x$. This then yields

$$f(\lambda x) \leq C x f(x), \quad \text{for} \quad x \leq x_0. \tag{5.43}$$

Iterating this equation, we get

$$f(\lambda^r x) \leq C^r \lambda^{r(r-1)/2} x^r f(x). \tag{5.44}$$

We can set $x = x_0$ in the above equation, giving

$$f(\lambda^r x_0) \leq C^r \lambda^{r(r-1)/2} x_0^r f(x_0). \tag{5.45}$$

Then taking $r \approx -\log x$ and rescaling the variables, we get

$$f(x) = \mathcal{O}(x^\alpha \exp[-\beta (\log x)^2]), \tag{5.46}$$

as $x \to 0$, where α and $\beta(> 0)$ are two constants dependent on λ. The gamma distribution decays slower than the right-hand side in Eq. (5.46) when $x \to 0$.

Thus the expression (5.46) gives an upper bound form at low wealth range and confirms again that the distribution of the global saving propensity model is not a gamma distribution.

5.1.3 The gas model analogy

The equilibrium distributions (5.11) can also be interpreted as the Gibbs distribution of the energy x, for a gas at temperature $T = \langle x \rangle / k_B$, which establishes a correspondence between the above models of closed economy and statistical systems, suggesting an interpretation of the economy model in terms of a mechanical system of interacting particles. Introducing a saving parameter $\lambda > 0$ changes the shape of the Gibbs distribution into that of a gamma distribution, but the correspondence with a mechanical system is lost only apparently. In fact, the Gibbs distribution (5.11) can represent the distribution of kinetic energy x only in $D = 2$ dimensions, when its average value is given by $\langle x \rangle_2 = 2(k_B T/2)$. In all other cases ($D \neq 2$), it is easy to show, starting from the Maxwell–Boltzmann distribution for the velocity in D dimensions, that the equilibrium kinetic energy distribution $f(x)$ coincides, except for a normalization factor, with the gamma distribution $\gamma_n(\xi)$ with $n = D/2$ for a reduced variable $\xi = Dx/2\langle x \rangle_D$,

$$\left.\begin{aligned} f(x) &= \frac{\left(\frac{D}{2\langle x \rangle_D}\right)^{D/2}}{\Gamma\left(\frac{D}{2}\right)} x^{D/2-1} \exp\left(-\frac{Dx}{2\langle x \rangle_D}\right), \\ \langle x \rangle_D &= D\langle x \rangle_1 = \frac{D k_B T}{2}, \end{aligned}\right\} \tag{5.47}$$

where $\langle x \rangle_D$ is the average value of kinetic energy in D dimensions. The analogy between the factor D in the argument of the exponential function in Eq. (5.47) and the analogous factor n in Eq. (5.7) is important. The main difference is that, while D is an integer number by hypothesis, the parameter $n(\lambda)$ can assume in general any real values larger than or equal to 1.

In Eq. (5.47) temperature appears implicitly as $T = 2\langle x \rangle_D / k_B D$. This suggests that also in the closed economy model considered above the effective temperature of the system should be defined as $\langle x \rangle / n$, rather than $\langle x \rangle$. This is a natural consequence of the fact that the average value of kinetic energy in D dimensions is proportional to D, owing to the equipartition theorem, and that an estimate of the amplitude of thermal fluctuations, which is independent of its effective dimension, can be obtained from the ratio $\langle x \rangle_D / D$. Direct comparison between Eq. (5.47) and Eq. (5.7) leads to a formal but exact analogy between money in the closed economy model considered above, with N agents, saving propensity $0 \leq \lambda \leq 1$ and given average money $\langle x \rangle$, on one hand, and kinetic energy in an ensemble of N particles

Table 5.1 *Analogy between the kinetic theory of gases and the kinetic exchange model of wealth*

	Kinetic model	Economy model
Variable	K (kinetic energy)	x (wealth)
Units	N particles	N agents
Interaction	Collisions	Trades
Dimension	Integer D	Real number D_λ
Temperature definition	$k_B T = 2\langle K\rangle/D$	$T_\lambda = 2\langle x\rangle/D_\lambda$
Reduced variable	$\xi = K/k_B T$	$\xi = x/T_\lambda$
Equilibrium distribution	$f(\xi) = \gamma_{D/2}(\xi)$	$f(\xi) = \gamma_{D_\lambda/2}(\xi)$

in D dimensions at temperature T, on the other, if the effective dimension and temperature are defined as

$$\left.\begin{aligned} D(\lambda) &= 2n(\lambda) = \frac{2(1+2\lambda)}{1-\lambda}, \\ T(\lambda) &= \frac{\langle x\rangle}{n(\lambda)} = \langle x\rangle\frac{1-\lambda}{1+2\lambda}, \end{aligned}\right\} \tag{5.48}$$

respectively. This equivalence can be qualitatively understood in terms of the underlying microscopic dynamics by considering the example of a fluid of interacting particles. In one dimension, particles undergo head-on collisions, in which they can exchange the total amount of energy they have. In an arbitrary (large) number of dimensions, however, this is not possible for purely kinematic reasons and only a fraction of the total energy is actually released or gained on average in a collision. Since the equipartition theorem implies that on average kinetic energy is equally shared among the D dimensions, one can expect that, during a collision, only a fraction $\sim 1/D$ of the total energy is exchanged (and that a corresponding fraction $\lambda \sim 1 - 1/D$ is 'saved'). This estimate of the exchanged energy $\sim 1/D$ is to be compared with the expression for the fraction of exchanged money obtained from Eq. (5.6) using $n = D/2$, namely $1 - \lambda = 3/(D/2 + 2)$, which was in fact found starting the fitting of the numerical data from a function prototype of a form similar to $1/D$. Thus, the ubiquitous presence of Γ-functions in the solutions of kinetic models (see also below, heterogeneous models) suggests a close analogy with the kinetic theory of gases. In fact, interpreting $D_\lambda = 2n$ as an effective dimension, the variable x as kinetic energy, and introducing the effective temperature $\beta^{-1} \equiv T_\lambda = \langle x\rangle/2D_\lambda$ according to the equipartition theorem, Eqs. (5.7) and (5.6) define the canonical distribution $\beta\gamma_n(\beta x)$ for the kinetic energy of a gas in $D_\lambda = 2n$ dimensions (see Patriarca *et al.* (2004) for details). The analogy is illustrated in Table 5.1 and the dependences of $D_\lambda = 2n$ and of $\beta^{-1} = T_\lambda$

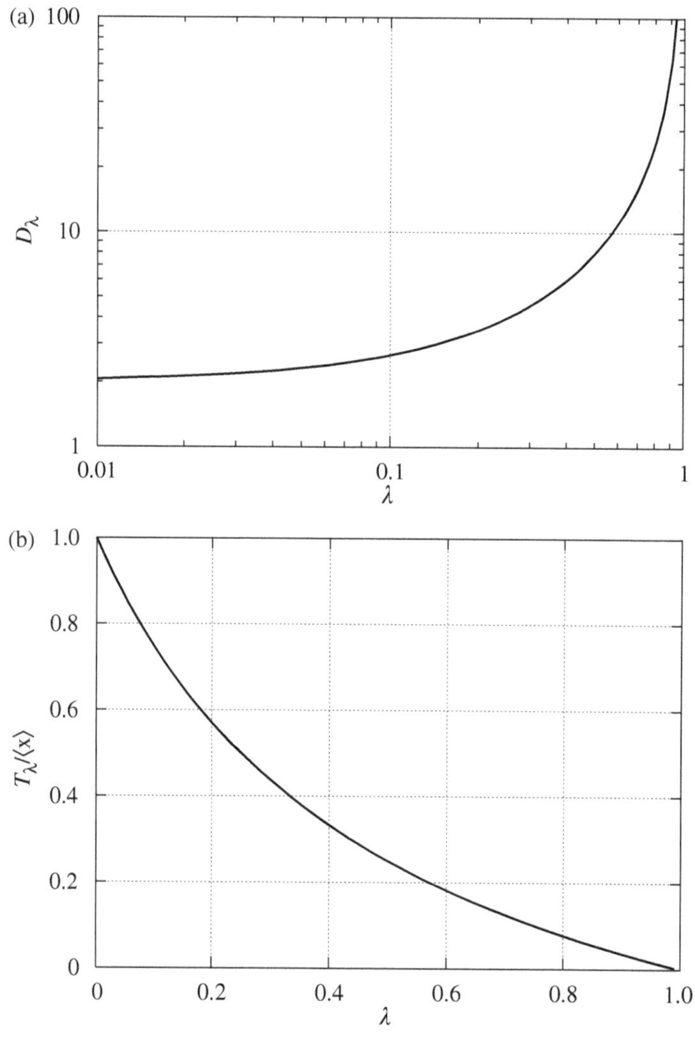

Figure 5.3 (a, b) Effective dimension D_λ and temperature T as a function of the saving parameter λ. Reproduced from Chakraborti and Patriarca (2008).

on the saving parameter λ are shown in Fig. 5.3. It can be shown that during a binary elastic collision in D dimensions only a fraction $1/D$ of the total kinetic energy is exchanged on average for kinematic reasons (see Chakraborti and Patriarca (2008) for details). The same $1/D$ dependence is in fact obtained inverting Eq. (5.6), which provides for the fraction of exchanged wealth $1 - \lambda = 6/(D_\lambda + 4)$.

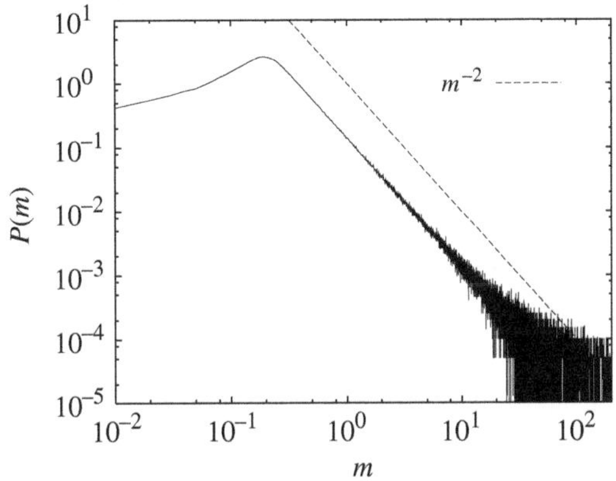

Figure 5.4 Steady-state money distribution $P(m)$ against m in a numerical simulation of a market with $N = 200$, following Eq. (4.13) and Eq. (4.14) with $\epsilon_{ij} = 1/2$. The dotted lines correspond to $m^{-(1+\nu)}$; $\nu = 1$. Here, the average money per agent $M/N = 1$. Reproduced from Chatterjee *et al.* (2005c).

5.2 Analytic results for the CCM model

The steady-state distribution of money, resulting from Eqs. (4.13) and (4.14) representing the money dynamics and trading, is investigated. The dynamics of money distribution are also analysed for two aspects. First, one studies the evolution of the mutual money difference among the agents and looks for a self-consistent equation for its steady-state distribution. Next, one can develop a master equation for the money distribution function (Chatterjee *et al.* 2005a,c).

5.2.1 Distribution of money difference

Clearly in the process as considered (dynamics defined by Eqs. (4.13) and (4.14)), the total money $(m_i + m_j)$ of the pair of agents i and j remains constant, while the difference Δm_{ij} evolves as

$$(\Delta m_{ij})_{t+1} \equiv (m_i - m_j)_{t+1} = \left(\frac{\lambda_i + \lambda_j}{2}\right)(\Delta m_{ij})_t + \left(\frac{\lambda_i - \lambda_j}{2}\right)(m_i + m_j)_t + (2\epsilon_{ij} - 1)[(1 - \lambda_i)m_i(t) + (1 - \lambda_j)m_j(t)]. \tag{5.49}$$

Numerically, as shown in Fig. 4.3, the steady-state money distribution in the market becomes a power law, following such tradings when the saving factor λ_i of the agents remains constant over time but varies from agent to agent widely. As shown in the numerical simulation results for $P(m)$ in Fig. 5.4, the law, as well as the exponent, remains unchanged even when $\epsilon_{ij} = 1/2$ for every trading. This can be justified by the earlier numerical observation (Chakraborti and Chakrabarti 2000; Chatterjee *et al.* 2004) for uniform λ market ($\lambda_i = \lambda$ for all i) that, in the steady state, criticality occurs as $\lambda \to 1$ where of course the dynamics become extremely slow. In other words, after the steady state is realized, the third term in Eq. (5.49) becomes unimportant for the critical behaviour. We therefore concentrate on this case, where the above evolution equation for Δm_{ij} can be written in a more simplified form as

$$(\Delta m_{ij})_{t+1} = \alpha_{ij}(\Delta m_{ij})_t + \beta_{ij}(m_i + m_j)_t, \tag{5.50}$$

where $\alpha_{ij} = \frac{1}{2}(\lambda_i + \lambda_j)$ and $\beta_{ij} = \frac{1}{2}(\lambda_i - \lambda_j)$. As such, $0 \leq \alpha < 1$ and $-\frac{1}{2} < \beta < \frac{1}{2}$.

The steady-state probability distribution D for the modulus $\Delta = |\Delta m|$ of the mutual money difference between any two agents in the market can be obtained from Eq. (5.50) in the following way provided Δ is very much larger than the average money per agent $= M/N$. This is because of the following reason: using Eq. (5.50), large Δ can appear at $t+1$, say, from 'scattering' from any situation at t for which the right-hand side of Eq. (5.50) is large. The possibilities are (at t) m_i large (rare) and m_j not large, where the right-hand side of Eq. (5.50) becomes $\sim(\alpha_{ij} + \beta_{ij})(\Delta_{ij})_t$; or m_j large (rare) and m_i not large (making the right-hand side of Eq. (5.50) become $\sim(\alpha_{ij} - \beta_{ij})(\Delta_{ij})_t$); or when m_i and m_j are both large, which is a much rarer situation than the first two and hence is negligible. Then if, say, m_i is large and m_j is not, the right-hand side of Eq. (5.50) becomes $\sim(\alpha_{ij} + \beta_{ij})(\Delta_{ij})_t$ and so on. Consequently, for large Δ, the distribution D satisfies

$$\begin{aligned} D(\Delta) &= \int d\Delta' \, D(\Delta') \, \langle \delta(\Delta - (\alpha+\beta)\Delta') + \delta(\Delta - (\alpha-\beta)\Delta') \rangle \\ &= 2\left\langle \left(\frac{1}{\lambda}\right) D\left(\frac{\Delta}{\lambda}\right) \right\rangle, \end{aligned} \tag{5.51}$$

where we have used the symmetry of the β distribution and the relation $\alpha_{ij} + \beta_{ij} = \lambda_i$, and have suppressed labels i, j. Here $\langle \ldots \rangle$ denotes average over λ distribution in the market. Taking now a uniform random distribution of the saving factor λ, $\rho(\lambda) = 1$ for $0 \leq \lambda < 1$, and assuming $D(\Delta) \sim \Delta^{-(1+\gamma)}$ for large Δ, we get

$$1 = 2\int d\lambda \, \lambda^{\gamma} = 2(1+\gamma)^{-1}, \tag{5.52}$$

giving $\gamma = 1$. No other value fits the above equation. This also indicates that the money distribution $P(m)$ in the market follows a similar power law variation, $P(m) \sim m^{-(1+\nu)}$ and $\nu = \gamma$. The distribution of Δ from numerical simulations also agrees with this result.

We will now show in a more rigorous way that indeed the only stable solution corresponds to $\nu = 1$, as observed numerically (Chakrabarti and Chatterjee 2004; Chatterjee *et al.* 2003, 2004).

5.2.2 Master equation and its analysis

We now proceed to develop a Boltzmann-like master equation (Chatterjee *et al.* 2005a,c) for the time development of $P(m, t)$, the probability distribution of money in the market. We again consider the case $\epsilon_{ij} = \frac{1}{2}$ in Eq. (4.13) and Eq. (4.14) and rewrite them as

$$\begin{pmatrix} m_i \\ m_j \end{pmatrix}_{t+1} = \mathcal{A} \begin{pmatrix} m_i \\ m_j \end{pmatrix}_t , \tag{5.53}$$

where

$$\mathcal{A} = \begin{pmatrix} \mu_i^+ & \mu_j^- \\ \mu_i^- & \mu_j^+ \end{pmatrix}; \quad \mu^\pm = \frac{1}{2}(1 \pm \lambda). \tag{5.54}$$

Collecting the contributions from terms scattering in and subtracting those scattering out, we can write the master equation for $P(m, t)$ as (see Slanina 2004)

$$\begin{aligned} P(m, t + \Delta t) - P(m, t) = & \left\langle \int dm_i \int dm_j \; P(m_i, t) P(m_j, t) \right. \\ & \times \{[\delta(\{\mathcal{A}\,\mathbf{m}\}_i - m) + \delta(\{\mathcal{A}\,\mathbf{m}\}_j - m)] \\ & \left. - [\delta(m_i - m) + \delta(m_j - m)]\} \right\rangle \\ = & \left\langle \int dm_i \int dm_j \; P(m_i, t) P(m_j, t) \right. \\ & \times [\delta(\mu_i^+ m_i + \mu_j^- m_j - m) + \delta(\mu_i^- m_i + \mu_j^+ m_j - m) \\ & \left. - \delta(m_i - m) + \delta(m_j - m)] \right\rangle . \end{aligned} \tag{5.55}$$

Here also, $\langle \ldots \rangle$ denotes the average over the distribution of λ.

The above equation can be rewritten as

$$\begin{aligned} & \frac{\partial P(m, t)}{\partial t} + P(m, t) \\ & = \left\langle \int dm_i \int dm_j \; P(m_i, t) P(m_j, t) \; \delta(\mu_i^+ m_i + \mu_j^- m_j - m) \right\rangle , \end{aligned} \tag{5.56}$$

which in the steady state gives

$$P(m) = \left\langle \int dm_i \int dm_j \ P(m_i)P(m_j)\ \delta(\mu_i^+ m_i + \mu_j^- m_j - m) \right\rangle. \qquad (5.57)$$

Writing $m_i \mu_i^+ = xm$, we can decompose the range [0, 1] of x into three regions: $[0, \kappa]$, $[\kappa, 1-\kappa']$ and $[1-\kappa', 1]$. Collecting the relevant terms in the three regions, we can rewrite the equation for $P(m)$ above as

$$\begin{aligned} P(m) &= \left\langle \frac{m}{\mu^+\mu^-} \int_0^1 dx\, P\left(\frac{xm}{\mu^+}\right) P\left(\frac{m(1-x)}{\mu^-}\right) \right\rangle \\ &= \left\langle \frac{m}{\mu^+\mu^-} \left\{ P\left(\frac{m}{\mu^-}\right) \frac{\mu^+}{m} \int_0^{\frac{\kappa m}{\mu^+}} dy P(y) + P\left(\frac{m}{\mu^+}\right) \frac{\mu^-}{m} \int_0^{\frac{\kappa' m}{\mu^-}} dy P(y) \right.\right. \\ &\quad \left.\left. + \int_\kappa^{1-\kappa'} dx\, P\left(\frac{xm}{\mu^+}\right) P\left(\frac{m(1-x)}{\mu^-}\right) \right\} \right\rangle, \qquad (5.58) \end{aligned}$$

where the result applies for κ and κ' sufficiently small. If we take $m \gg 1/\kappa$, $m \gg 1/\kappa'$ and $\kappa, \kappa' \to 0$ $(m \to \infty)$, then

$$\begin{aligned} P(m) &= \left\langle \frac{m}{\mu^+\mu^-} \left\{ P\left(\frac{m}{\mu^-}\right) \frac{\mu^+}{m} + P\left(\frac{m}{\mu^+}\right) \frac{\mu^-}{m} \right.\right. \\ &\quad \left.\left. + \int_\kappa^{1-\kappa'} dx\, P\left(\frac{xm}{\mu^+}\right) P\left(\frac{m(1-x)}{\mu^-}\right) \right\} \right\rangle. \qquad (5.59) \end{aligned}$$

Assuming now as before, $P(m) = A/m^{1+\nu}$ for $m \to \infty$, we get

$$1 = \langle (\mu^+)^\nu + (\mu^-)^\nu \rangle \equiv \int\int d\mu^+ d\mu^- p(\mu^+) q(\mu^-) [(\mu^+)^\nu + (\mu^-)^\nu], \qquad (5.60)$$

as the ratio of the third term in Eq. (5.59) to the other terms vanishes like $(m\kappa)^{-\nu}$, $(m\kappa')^{-\nu}$ in this limit and $p(\mu^+)$ and $q(\mu^-)$ are the distributions of the variables μ^+ and μ^-, which vary uniformly in the ranges $[\frac{1}{2}, 1]$ and $[0, \frac{1}{2}]$, respectively (cf. Eq. (5.54)). The i, j indices, for μ^+ and μ^-, are again suppressed here in Eq. (5.60), and we utilize the fact that μ_i^+ and μ_j^- are independent for $i \neq j$. An alternative way of deriving Eq. (5.60) from Eq. (5.57) is to consider the dominant terms ($\propto x^{-r}$ for $r > 0$, or $\propto \ln(1/x)$ for $r = 0$) in the $x \to 0$ limit of the integral $\int_0^\infty m^{(\nu+r)} P(m) \exp(-mx) dm$ (see next subsection). We therefore get from Eq. (5.60), after integrations, $1 = 2/(\nu + 1)$, giving $\nu = 1$.

5.2.2.1 Alternative solution of the steady-state master equation

Let $S_r(x) = \int_0^\infty dm\, P(m) m^{\nu+r} \exp(-mx)$; $r \geq 0$, $x > 0$. If $P(m) = A/m^{1+\nu}$, then

$$
\begin{aligned}
S_r(x) &= A \int_0^\infty dm\ m^{r-1} \exp(-mx) \\
&\sim A\,\frac{x^{-r}}{r} \qquad \text{if } r > 0 \\
&\sim A \ln\left(\frac{1}{x}\right) \qquad \text{if } r = 0.
\end{aligned}
\tag{5.61}
$$

From Eq. (5.57), we can write

$$
\begin{aligned}
S_r(x) = &\left\langle \int_0^\infty dm_i \int_0^\infty dm_j\ P(m_i)P(m_j)(m_i\mu_i^+ + m_j\mu_j^-)^{\nu+r} \right. \\
&\left. \times \exp[-(m_i\mu_i^+ + m_j\mu_j^-)x] \right\rangle \\
\simeq &\int_0^\infty dm_i\ Am_i^{r-1}\langle \exp(-m_i\mu_i^+ x)(\mu_i^+)^{\nu+r}\rangle \\
&\times \left[\int_0^\infty dm_j\ P(m_j)\langle \exp(-m_j\mu_j^- x)\rangle\right] \\
&+ \int_0^\infty dm_j\ Am_j^{r-1}\langle \exp(-m_j\mu_j^- x)(\mu_j^-)^{\nu+r}\rangle \\
&\times \left[\int_0^\infty dm_i\ P(m_i)\langle \exp(-m_i\mu_i^+ x)\rangle\right],
\end{aligned}
\tag{5.62}
$$

or

$$
\begin{aligned}
S_r(x) = &\int_{\frac{1}{2}}^1 d\mu_i^+\ p(\mu_i^+)\left(\int_0^\infty dm_i\ Am_i^{r-1}\exp(-m_i\mu_i^+ x)\right)(\mu_i^+)^{\nu+r} \\
&+ \int_0^{\frac{1}{2}} d\mu_j^-\ q(\mu_j^-)\left(\int_0^\infty dm_j\ Am_j^{r-1}\exp(-m_j\mu_j^- x)\right)(\mu_j^-)^{\nu+r},
\end{aligned}
\tag{5.63}
$$

since, for small x, the terms in the square brackets in Eq. (5.62) approach unity. We can therefore rewrite Eq. (5.63) as

$$
S_r(x) = 2\left[\int_{\frac{1}{2}}^1 d\mu^+(\mu^+)^{\nu+r} S_r(x\mu^+) + \int_0^{\frac{1}{2}} d\mu^-(\mu^-)^{\nu+r} S_r(x\mu^-)\right].
\tag{5.64}
$$

Using now the forms of $S_r(x)$ as in Eq. (5.61), and collecting terms of order x^{-r} (for $r > 0$) or of order $\ln(1/x)$ (for $r = 0$) from both sides of Eq. (5.64), we get Eq. (5.60).

Employing a mean-field approach, Mohanty (2006) calculated the distribution for the ensemble average of money for the model with distributed savings. It is assumed that the distribution of money of a single agent over time is stationary, which means that the time-averaged value of money of any agent remains unchanged independent of the initial value of money. Taking the ensemble average of all terms on both sides of Eq. 4.13, one can write

$$\langle m_i \rangle = \lambda_i \langle m_i \rangle + \langle \epsilon \rangle \left[(1-\lambda_i)\langle m_i \rangle + \left\langle \frac{1}{N} \sum_{j=1}^{N} (1-\lambda_j) m_j \right\rangle \right]. \tag{5.65}$$

The last term on the right is replaced by the average over the agents, where it is assumed that any agent (i-th agent here), on the average, interacts with all others in the system, which is the *mean-field* approach.

Writing

$$\overline{\langle (1-\lambda) m \rangle} \equiv \left\langle \frac{1}{N} \sum_{j=1}^{N} (1-\lambda_j) m_j \right\rangle \tag{5.66}$$

and since ϵ is assumed to be distributed randomly and uniformly in [0, 1], so that $\langle \epsilon \rangle = 1/2$, Eq. 5.65 reduces to

$$(1-\lambda_i)\langle m_i \rangle = \overline{\langle (1-\lambda) m \rangle}. \tag{5.67}$$

Since the right-hand side is free of any agent index, it seems that this relation is true for any arbitrary agent, i.e. $\langle m_i \rangle (1-\lambda) =$ constant, where λ is the saving factor of the i-th agent. What follows is $d\lambda =$ const. $\frac{dm}{m^2}$. An agent with a (characteristic) saving propensity factor (λ) ends up with wealth (m) such that one can in general relate the distributions of the two:

$$P(m)\, dm = \rho(\lambda)\, d\lambda. \tag{5.68}$$

Therefore, the distribution in m is bound to be of the form:

$$P(m) \propto \frac{1}{m^2}$$

for uniform distribution of savings factor λ, i.e. $\nu = 1$. This analysis can also explain the non-universal behaviour of the Pareto exponent ν, i.e. $\nu = 1 + \alpha$ for $\rho(\lambda) = (1-\lambda)^\alpha$. Thus, this mean-field study explains the origin of the universal ($\nu = 1$) and the non-universal ($\nu \neq 1$) Pareto exponents in the distributed savings model (Kar Gupta 2006a; Mohanty 2006).

5.2.3 A transition matrix formalism

To understand the qualitative differences in the various types of exchange processes in the above models, one can also look at the matrix $\mathcal{M}$ (Eq. 4.1) and its properties (Kar Gupta 2006a,b). For example, in the case of the pure gambling process (Drăgulescu and Yakovenko 2000)

$$\mathcal{M} = \begin{pmatrix} \alpha & \alpha \\ 1-\alpha & 1-\alpha \end{pmatrix}. \tag{5.69}$$

The above matrix is *singular* (determinant, $|\mathcal{M}| = 0$), which means that the inverse of this matrix does not exist. This indicates that an evolution through such transition matrices is bound to be *irreversible*. This property is connected to the emergence of an exponential (Boltzmann–Gibbs) wealth distribution. The same may be perceived in a different way too. When a product of such matrices (for successive interactions) is taken, the left-most matrix (of the product) itself returns:

$$\begin{pmatrix} \alpha & \alpha \\ 1-\alpha & 1-\alpha \end{pmatrix}\begin{pmatrix} \alpha_1 & \alpha_1 \\ 1-\alpha_1 & 1-\alpha_1 \end{pmatrix} = \begin{pmatrix} \alpha & \alpha \\ 1-\alpha & 1-\alpha \end{pmatrix}. \tag{5.70}$$

The above indicates the fact that, during the repeated interactions of the same pair of agents (through this kind of transition matrices), the last of the interactions is what matters (the last matrix of the product survives)

$$\mathcal{M}^{(n)}.\mathcal{M}^{(n-1)}\ldots\mathcal{M}^{(2)}.\mathcal{M}^{(1)} = \mathcal{M}^{(n)}.$$

This 'loss of memory' (random history of collisions in the case of molecules) may be attributed here to the path to irreversibility in time.

One can consider the following general form:

$$\mathcal{M}_1 = \begin{pmatrix} \alpha_1 & \alpha_2 \\ 1-\alpha_1 & 1-\alpha_2 \end{pmatrix}, \tag{5.71}$$

where α_1 and α_2 are two different random numbers drawn uniformly from [0, 1]. This ensures that the transition matrix is non-singular. The significance of this can be seen through the wealth exchange equations in the following manner: α_1 fraction of wealth of the first agent i added with α_2 fraction of wealth of the second agent j is retained by the first agent after the transaction. The rest of their total wealth is shared by the second agent. This may happen in several ways which can be attributed to the details of a model. The general matrix $\mathcal{M}_1$ is non-singular as long as $\alpha_1 \neq \alpha_2$ and then the two-agent interaction process remains reversible in time. Hence, it is expected to have a steady-state equilibrium distribution of wealth, which could have a form different from the exponential distribution (as in the pure gambling model); $\alpha_1 = \alpha_2$ gives back the pure exponential distribution (Kar Gupta

2006a,b). One can check with a trivial case for $\alpha_1 = 1$ and $\alpha_2 = 0$. The transition matrix reduces to the identity matrix $I = \begin{pmatrix} 1 & 0 \\ 0 & 1 \end{pmatrix}$, which trivially corresponds to no evolution.

It was emphasized that any transition matrix $\begin{pmatrix} t_{11} & t_{12} \\ t_{21} & t_{22} \end{pmatrix}$ for such conserved models is bound to be of the form such that the sum of two elements of either of the columns has to be *unity by design*: $t_{11} + t_{21} = 1$, $t_{12} + t_{22} = 1$. It is important to note that, whatever extra parameter one incorporates within the framework of the conserved model, the transition matrix has to retain this property.

One can easily play around with a combination of different parameters, and see that one can produce a wide range of distributions. An easy check with the CC model (Chakraborti and Chakrabarti 2000) can demonstrate the usefulness of the formulation. The matrix $\mathcal{M}$ looks like

$$\begin{pmatrix} \lambda + \epsilon(1-\lambda) & \epsilon(1-\lambda) \\ (1-\epsilon)(1-\lambda) & \lambda + (1-\epsilon)(1-\lambda) \end{pmatrix}. \tag{5.72}$$

One can do a rescaling $\tilde{\alpha}_1 = \lambda + \epsilon(1-\lambda)$ and $\tilde{\alpha}_2 = \epsilon(1-\lambda)$, which reduces the above transition matrix to

$$\mathcal{M}_2 = \begin{pmatrix} \tilde{\alpha}_1 & \tilde{\alpha}_2 \\ 1-\tilde{\alpha}_1 & 1-\tilde{\alpha}_2 \end{pmatrix}, \tag{5.73}$$

which has the same form as $\mathcal{M}_\infty$. The distributions due to the above two matrices can be compared if one correctly identifies the ranges of the rescaled elements. In the model of uniform saving: $\lambda < \tilde{\alpha}_1 < 1$ and $0 < \tilde{\alpha}_2 < (1-\lambda)$ as the stochasticity parameter ϵ is drawn from a uniform and random distribution in [0, 1]. As long as $\tilde{\alpha}_1$ and $\tilde{\alpha}_2$ are different, the determinant of the matrix is non-zero ($|\mathcal{M}_2| = \tilde{\alpha}_1 - \tilde{\alpha}_2 = \lambda$). Thus, the incorporation of the saving propensity factor λ makes: (1) the transition matrix non-singular and (2) the matrix elements $t_{11} (= \tilde{\alpha}_1)$ and t_{12} $(= \tilde{\alpha}_2)$ are now drawn from truncated domains (somewhere in [0, 1]).

The above makes it clear that the wealth distribution with uniform saving is likely to be qualitatively similar to what can be produced with general transition matrices, but having different elements, $\alpha_1 \neq \alpha_2$. The distributions obtained with different λ may correspond to those with appropriately chosen α_1 and α_2 in $\mathcal{M}_1$ (Kar Gupta 2006a,b).

For the CCM model (Chatterjee *et al.* 2004), the matrix looks like

$$\begin{pmatrix} \lambda_1 + \epsilon(1-\lambda_1) & \epsilon(1-\lambda_2) \\ (1-\epsilon)(1-\lambda_1) & \lambda_2 + (1-\epsilon)(1-\lambda_2) \end{pmatrix}, \tag{5.74}$$

where the elements can be rescaled by putting $\tilde{\alpha}_1' = \lambda_1 + \epsilon(1-\lambda_1)$ and $\tilde{\alpha}_2' = \epsilon(1-\lambda_2)$. Hence the transition matrix can again be reduced to the same form as that of $\mathcal{M}_1$ or $\mathcal{M}_2$:

$$\mathcal{M}_3 = \begin{pmatrix} \tilde{\alpha}'_1 & \tilde{\alpha}'_2 \\ 1-\tilde{\alpha}'_1 & 1-\tilde{\alpha}'_2 \end{pmatrix}. \tag{5.75}$$

The determinant here is $|\mathcal{M}_3| = \tilde{\alpha}'_1 - \tilde{\alpha}'_2 = \lambda_1(1-\epsilon) + \epsilon\lambda_2$. Here also the determinant is ensured to be non-zero as all the parameters ϵ, λ_1 and λ_2 are drawn from the same positive domain: [0, 1]. This means that each transition matrix for two-agent wealth exchange remains non-singular, which ensures that the interaction process is reversible in time. Thus, one can expect qualitatively different distributions by appropriately tuning two independent elements in the transition matrices (Eqs. (5.71), (5.73) or (5.75)). It is also known that, in the framework of the CCM model, the saving propensity has to have a broad, quenched distribution to produce a power law in money distribution, and this is equivalent to a reduced situation modelled with a single parameter η (Kar Gupta 2006a). The matrix looks like

$$\mathcal{M}_4 = \begin{pmatrix} 1 & \eta \\ 0 & 1-\eta \end{pmatrix}, \tag{5.76}$$

and it has a non-zero determinant ($|\mathcal{M}_4| = 1-\eta \neq 0$).

Thus, one can reformulate the above cases with a general form of the matrix

$$\begin{pmatrix} \alpha_1 & \alpha_2 \\ 1-\alpha_1 & 1-\alpha_2 \end{pmatrix}, \tag{5.77}$$

and one can generate all sorts of distributions by controlling α_1 and α_2.

As long as $\alpha_1 \neq \alpha_2$, the matrix remains non-singular and it results in gamma-type distributions. When $\alpha_1 = \alpha_2$, the transition matrix becomes singular and it produces a Boltzmann–Gibbs-type exponential distribution. The power law with exponent $\nu = 1$ is obtained with the general matrix when the elements α_1 and α_2 are of the same set of quenched random numbers drawn uniformly in [0, 1]. The matrix corresponding to the reduced situation as discussed (5.76) is a special case with $\alpha_1 = 1$ and $\alpha_2 = \eta$, drawn from a uniform and (quenched) random distribution. Incorporation of any parameter in an actual model (say, saving propensity) results in an adjustment or truncation of the full domain [0, 1] from which the element α_1 or α_2 is drawn. Incorporating distributed λ's is equivalent to considering the following domains: $\lambda_1 < \alpha_1 < 1$ and $0 < \alpha_2 < (1-\lambda_2)$.

5.2.4 Entropy maximization in a heterogeneous system

First, we review the results for a dimensionally homogeneous system (Patriarca *et al.* 2004). For a quadratic energy function $x(\mathbf{q}) = (q_1^2 + \cdots + q_N^2)/2$, where the q_i's are the N variables, for example, of a polymer, the equilibrium density is the functional $S_n[f_n]$, obtained from the Boltzmann entropy (Patriarca *et al.* 2004),

$$S_n[f_n] = \int_0^{+\infty} dx\ f_n(x) \left\{ \ln\left[\frac{f(x)}{\sigma_{2n}\, x^{n-1}}\right] + \mu + \beta x \right\}, \tag{5.78}$$

where one introduced the dimension variable

$$n = N/2\,, \tag{5.79}$$

σ_N is the hypersurface of a unitary N-dimensional sphere, and μ, β are Lagrange multipliers determined by the constraints on the conservation of the total number of subsystems and energy, respectively. The result is a Γ-distribution of order n,

$$f_n(x) = \beta\, \gamma_n(\beta x) \equiv \frac{\beta}{\Gamma(n)}(\beta x)^{n-1} \exp(-\beta x)\,, \tag{5.80}$$

where $\beta^{-1} = 2\langle x\rangle/N = \langle x\rangle/n$, according to the equipartition theorem, is the temperature (Patriarca *et al.* 2004).

Chakraborti and Patriarca (2009) suggested that a heterogeneous system composed of agents with different λ_i (CCM model) is analogous to a dimensionally heterogeneous system. In the case of a uniform distribution for the saving parameters, $\phi(\lambda) = 1$ if $\lambda \in (0, 1)$ and $\phi(\lambda) = 0$ otherwise, setting $n = N/2$, the dimension density has a power law $\sim 1/n^2$, $P(n) = \phi(\lambda)d\lambda/dn = 3/(n+2)^2$ $(n \geq 1)$. Thus, the heterogeneous system is characterized according to a distribution $P(n)$, with $\int dn\ P(n) = 1$. At equilibrium, each subsystem with dimension variable n will have its probability density $f_n(x)$. One is interested in the shape of the aggregate equilibrium energy distribution, i.e. the relative probability of finding a subsystem with energy x independently of its dimension n. The equilibrium problem for the heterogeneous system is solved analogously, from the functional $S[\{f_n\}]$ obtained summing the homogeneous functionals with different n,

$$S[\{f_n\}] = \int dn\ P(n) \int_0^{+\infty} dx\ f_n(x) \left\{ \ln\left[\frac{f_n(x)}{\sigma_{2n}\, x^{n-1}}\right] + \mu_n + \beta x \right\}. \tag{5.81}$$

Notice that there is a different Lagrange multiplier μ_n for each n, since the fractions $P(n)$ are conserved separately, but a single β is related to the total conserved energy. The equilibrium probability density $f_n(x)$ for the subsystem n is obtained by varying $S[\{f_n\}]$ with respect to $f_n(x)$ and is again given by Eq. (5.80), with β determined by the total energy,

$$\langle x\rangle = \int dn \int_0^{\infty} dx\ f_n(x)\, x = \frac{\langle N\rangle}{2\beta}\,, \tag{5.82}$$

where we have introduced the average dimension

$$\langle N\rangle = 2\langle n\rangle = 2\int dn\ P(n)\, n\,. \tag{5.83}$$

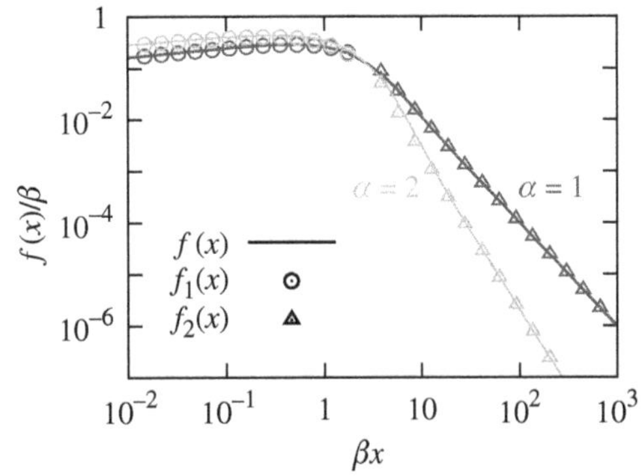

Figure 5.5 Aggregate distribution $f(x)$, Eq. (5.84), with $P(n) = \alpha/n^{1+\alpha}$ $(n \geq 1)$, $P(n) = 0$ otherwise, for $\alpha = 1$ (dark grey), $\alpha = 2$ (light grey). Continuous lines: numerical integration of Eq. (5.84). Triangles: saddle point approximation $f_2(x)$, Eq. (5.91). Circles: small-x limit, $f_1(x)$, Eq. (5.92). Reproduced from Chakraborti and Patriarca (2009).

Equation (5.82) represents a *generalized equipartition theorem* for dimensionally heterogeneous systems. To ensure a finite $\langle N \rangle$ (and therefore a finite average energy $\langle x \rangle$) the dimension density $P(n)$ has to have a finite cut-off or decrease faster than $1/n^2$ for $n \gg 1$. Using Eq. (5.80), the aggregate distribution is finally

$$f(x) = \int dn\, P(n) f_n(x) = \int dn\, \frac{P(n)\beta}{\Gamma(n)} (\beta x)^{n-1} \exp(-\beta x). \qquad (5.84)$$

While the distributions $f_n(x)$ have an exponential tail, the function $f(x)$ may exhibit a slower decay and possibly a power law tail, if the dimension density $P(n)$ decreases slowly enough. In the example of the power law density $P_\alpha(n) = \alpha/n^{1+\alpha}$ $(n \geq 1, \alpha > 0)$, $P_\alpha(n) = 0$ otherwise, a power law tail appears in $f(x)$; see the continuous curves in Fig. 5.5 obtained by numerical integration.

In fact, a general result holds, namely an actual equivalence between the asymptotic form of the aggregate distribution $f(x)$ and the dimension density $P(n)$, whenever $P(n)$ decreases at large n faster than $1/n$, expressed by $f(x \gg \beta^{-1}) = \beta P(\beta x)$. This asymptotic relation can be compared with the equality between the the average load and the degree distribution, $f(x) = g(x/\bar{x})/\bar{x}$, obtained for the example of free diffusion on a network with degree distribution $g(k)$.

To demonstrate this relation, we start by considering a value $\beta x \gg 1$ in Eq. (5.84). The main contributions to the integral come from values $n \approx \beta x \gg 1$, since $\gamma_n(\beta x)$ has its maximum at $x \approx n/\beta$, while it goes to zero for small as well as larger x. Introducing the variable $m = n - 1$, Eq. (5.84) can be rewritten as

$$f(x) = \beta \exp(-\beta x) \int dm \, \exp[-\phi(m)] , \tag{5.85}$$

$$\phi(m) = -\ln[P(m+1)] - m \ln(\beta x) + \ln[\Gamma(m+1)] . \tag{5.86}$$

This integral can be estimated through the saddle-point approximation expanding $\phi(m)$ to the second order in $\epsilon = m - m_0$, where $m_0 = m_0(x)$ locates the maximum of $\phi(m)$, defined by $\phi'(m_0) = 0$ and $\phi''(m_0) > 0$, and integrating over the whole range of m,

$$f(x) \approx \beta \exp[-\beta x - \phi(m_0)] \int_{-\infty}^{+\infty} d\epsilon \exp[-\phi''(m_0)\epsilon^2/2]$$

$$= \beta \sqrt{\frac{2\pi}{\phi''(m_0)}} \exp[-\beta x - \phi(m_0)] . \tag{5.87}$$

In order to find m_0 we use the Stirling approximation (Abramowitz and Stegun 1970) in Eq. (5.86), $\Gamma(m+1) \approx \sqrt{2\pi m}\,(m/e)^m$, so that

$$\phi(m) \approx -\ln[P(m+1)] - m \ln(\beta x) + \ln(\sqrt{2\pi}) + \left(m + \frac{1}{2}\right) \ln(m) - m, \tag{5.88}$$

$$\phi'(n) \approx -\frac{P'(m+1)}{P(m+1)} - \ln(\beta x) + \frac{1}{2m} + \ln(m) , \tag{5.89}$$

$$\phi''(n) \approx \frac{P'^2(m+1)}{P^2(m+1)} - \frac{P''(m+1)}{P(m+1)} - \frac{1}{2m^2} + \frac{1}{m} . \tag{5.90}$$

From Eq. (5.90) the condition $\phi''(m) > 0$ for the existence of a maximum is fulfilled for large m, if one can neglect the terms containing P with respect to $1/m$; this can be done for general shapes of $P(n)$ which decrease fast enough.

From Eq. (5.89) in the same limit one can neglect P'/P and $1/m$ with respect to $\ln(m)$ and the approximate solution of $\phi'(m_0) = 0$ is $m_0(x) \approx \beta x$. It can be checked that even keeping higher orders in $1/m$ in Eq. (5.89) the asymptotic solution reduces to $m_0(x) = \beta x$ for $\beta x \gg 1$. Finally, setting $m = m_0(x) = \beta x$ and using Eqs. (5.88) and (5.90) in Eq. (5.87), one finds

$$f(x \gg \beta^{-1}) \equiv f_2(x) = \beta P(1 + \beta x) . \tag{5.91}$$

This relation provides the asymptotic form of the density $f(x)$ directly in terms of the dimension density $P(n)$, in the hypothesis that $P(n)$ decreases with n at least as $1/n$.

The approximate form of $f(x)$ at $x \ll \beta^{-1}$ depends on the details of $P(n)$ at small n. For the same form $P(n) = P_\alpha(n)$ considered above and setting $\phi(n) \approx \phi(1) + \phi'(1)(n-1)$ in Eq. (5.85) and Eq. (5.86), one has

$$f(x \ll \beta^{-1}) \equiv f_1(x) = \beta \int_0^{+\infty} dm \exp[-\phi(0) - \phi'(0)m - \beta x]$$
$$= \frac{\beta P(1) \exp(-\beta x)}{-\ln(\beta x) - \gamma - P'(1)/P(1)}. \tag{5.92}$$

Here, from Eq. (5.86), we set $\phi(0) = \ln[P(1)]$ and $\phi'(0) = -\gamma - \ln(\beta x) - P'(1)/P(1)$, with $\gamma = \psi(1) \equiv (d \ln[\Gamma(m)]/dm)_{m=1} \approx 0.57721$ the Euler γ-constant (Abramowitz and Stegun 1970).

In Fig. 5.5 the function $f_2(x)$ (triangles), given by Eq. (5.91), is compared at large x with the exact distribution $f(x)$ obtained by numerical integration of Eq. (5.84) (continuous lines) for the values $\alpha = 1, 2$ for the power law density $P_\alpha(n)$. Also the corresponding density $f_1(x)$ (circles), given by Eq. (5.92), is shown at small βx.

Thus, Chakraborti and Patriarca (2009) showed that in a conservative mechanical system composed of subsystems with different numbers of degrees of freedom a robust power law tail can appear in the equilibrium distribution of energy as a result of certain superpositions of the canonical equilibrium energy densities of the subsystems. The derivation only used the variational principle based on the Boltzmann entropy, without assumptions outside the framework of canonical equilibrium statistical mechanics.

5.3 Analytic results for other models

5.3.1 Autoregressive models and non-conservative form

Basu and Mohanty (2008) had put forward an interesting formalism to treat models of similar kind, and even loosening the conservation constraint to go beyond the microcanonical picture. They consider N independent agents $i = 1 \ldots N$, whose wealth at a given time t is $x_i(t)$. As before, each agent i invests a definite fraction of wealth $\mu_i x_i(t)$ in the market, which stochastically returns a net gain $\xi(t)$. Thus, the wealth of agent i at time t is

$$x_i(t) = (1 - \mu_i) x_i(t-1) + \xi(t), \tag{5.93}$$

where $\xi(t)$ is an uncorrelated positive stochastic variable with probability distribution function (PDF) $h(\xi)$ and does not depend on $\{x_i\}$. Thus, agents may gain or lose from the market. The autoregressive nature of the model that $x(t)$ depends on $x(t-1)$ is clear from Eq. (5.93).

To look for the steady state of Eq. (5.93), one can define an operator $\mathcal{B}$ which takes the variables one step backward in time, i.e. $x(t-1) = \mathcal{B}x(t)$. For convenience let $\lambda_i \equiv 1 - \mu_i$, similar to the savings propensity defined in kinetic models (Chakraborti and Chakrabarti 2000). Now, Eq. (5.93) becomes

$$x(t) = \frac{1}{1-\lambda\mathcal{B}}\xi(t) = \sum_{n=0}^{\infty} \lambda^n \xi(t-n) = \sum_{n=0}^{t} \lambda^n \xi(n), \tag{5.94}$$

where the index i is dropped because the agents are independent. One also uses the fact that $\xi(t)$ is an uncorrelated random variable and that $\xi(n<0) = 0$. Thus, the steady-state distribution $P(x)$ which is reached as $t \to \infty$ is the PDF of the stochastic variable

$$x = \sum_{n=0}^{\infty} \lambda^n \xi(n), \tag{5.95}$$

which is the weighted sum of $\{\xi(n)\}$ having weights$\{\lambda^n\}$. Let $x_m = \sum_{n=0}^{m} \lambda^n \xi(n)$ be the first m terms of (5.95) and their distribution be $P_m(x)$. From Eq. (5.94) and Eq. (5.95) it follows that $x_m = x(t=m)$. First, it means that the true steady state gets contributions from all orders of λ^n. Second, $P_m(x)$ can be considered as the distribution at $t = m$.

Since $x_m = \lambda^m \xi(m) + x_{m-1}$, $P_m(x)$ satisfies a recursion relation,

$$P_m(x) = \frac{1}{\lambda^m} \int_0^x P_{m-1}(y) h\left(\frac{x-y}{\lambda^m}\right) dy. \tag{5.96}$$

The steady-state distribution becomes $P(x) \equiv \lim_{m\to\infty} P_m(x)$. Clearly, from Eq. (5.96), it follows that

$$P_m(0) = 0 \quad \text{for all} \quad m > 0. \tag{5.97}$$

Thus, in the steady state one must have $P(x=0) = 0$. Equation (5.97), being independent of the choice of $h(\xi)$, can be used as the general boundary condition for Eq. (5.96). Additionally, it indicates that the steady-state distribution is neither Gibbs nor pure power law like, where $P(x=0)$ is finite.

At this stage one requires the information about the specific form of the function $h(\xi)$. For static markets, where the average wealth of the market is fixed, $a \equiv \langle \xi \rangle$. Basu and Mohanty (2008) finally discuss the conditions under which the different types of distributions arise, and also under different conditions of markets.

(1) *Normal distribution of* ξ. When the fluctuation of the market is normal, i.e. $h(\xi)$ is a Gaussian distribution denoted by $\mathcal{G}(\alpha_0, \sigma_0)$ with mean α_0 and standard

deviation σ_0, the steady-state distribution $P(x)$ is $\mathcal{G}(\alpha, \sigma)$, where

$$\alpha = \frac{\alpha_0}{1-\lambda} \quad \text{and} \quad \sigma = \frac{\sigma_0}{\sqrt{1-\lambda^2}}. \tag{5.98}$$

$\mathcal{G}(\alpha, \sigma)$ satisfies Eq. (5.93) in the steady state; i.e. if the PDFs of x and ξ are $\mathcal{G}(\mu_0, \sigma_0)$ and $\mathcal{G}(\mu, \sigma)$, respectively, then the PDF of $\lambda x + \xi$ is the same as that of x. Note that agents in this case can have negative wealth even though $\langle x \rangle > 0$. The negative wealth can be interpreted as *debt*.

(2) *Exponential distribution of* ξ. Let $h(\xi) = \exp(-\xi)$. For $\lambda = 0$ it gives the same steady-state distribution as that of the CC model (Chakraborti and Chakrabarti 2000), i.e. $P(x) \propto \exp(-x)$. For $\lambda \neq 0$, one needs to solve the integral equation (5.96). Instead it can be rewritten as a differential equation

$$\frac{d}{dx} P_m(x) = \frac{1}{\lambda^m} \left[P_{m-1}(x) - P_m(x) \right], \tag{5.99}$$

where $m > 0$, and the boundary conditions are given by Eq. (5.97). For $m = 0$, $P_0(x) \equiv h(x)$. In terms of $G_m(s)$, which is the Laplace transform of $P_m(x)$, Eq. (5.99) becomes a difference equation

$$G_m(s) = \frac{1}{1+\lambda^m s} G_{m-1}(s), \tag{5.100}$$

whose formal solution is

$$G_m(s) = \prod_{k=0}^{m-1} (1 + \lambda^k s)^{-1} G_0(s).$$

Again, remembering that $G_0(s)$ is the Laplace transform of $P_0(x) = h(x)$. Finally, $P(x)$ is the inverse Laplace transform of

$$G(s) = \prod_{k=1}^{\infty} \frac{1}{1+\lambda^k s} G_0(s), \tag{5.101}$$

which can be written as the following series:

$$\left. \begin{aligned} & P(x) = \sum_{m=1}^{\infty} C_m \exp(-x/\lambda^m), \\ & \text{where } C_m^{-1} = \lambda^m \prod_{0<n\neq m}^{\infty} (1 - \lambda^{n-m}). \end{aligned} \right\} \tag{5.102}$$

Equation (5.102) is an infinite series, but the first few terms are good enough for numerical evaluation of the distribution. Terms up to $m = n$ give $P_n(x)$, which can be interpreted either as an approximation of the true steady-state distribution $P(x)$ to n-th order in λ or as the distribution at finite time $t = n$. Numerically, it is found that $P(x)$ is a gamma-like distribution similar to what has been obtained in Ispolatov *et al.* (1998), Chakraborti and Chakrabarti (2000) and Bhattacharya *et al.* (2005).

(3) *Pareto-law*. In the above consideration, the wealth distributions of individual agents are not simple and depend on their investment capacities μ_i. Their averages, however, follow a power law. To prove this one has to define $\langle x_i \rangle = w_i$. In steady state $\langle x(t) \rangle = \langle x(t-1) \rangle$. Thus, Eq. (5.93) gives

$$w_i = \frac{\langle \xi \rangle}{\mu_i}, \tag{5.103}$$

with agents differing in their investment capacities. The average number of agents having investment capacity μ is $Ng(\mu)$, where $g(\mu)$ is the distribution of μ. Thus, one can write $w(\mu) = \langle \xi \rangle / \mu$. Distribution of w is then

$$P(w) = g(\mu)|\frac{d\mu}{dw}| = \langle \xi \rangle \frac{g(\langle \xi \rangle / w)}{w^2}, \tag{5.104}$$

similar to the argument used in Mohanty (2006) for deriving the wealth distribution of the CCM model. Although distribution for the rich (large w) is generically $P(w) \sim w^{-2}$, one can obtain $\gamma > 2$ in typical cases. For example, if the PDF of μ is $g(\mu) = \mu^{\alpha}/(\alpha - 1)$ with $0 \leq \alpha < 1$, the asymptotic distribution of Eq. (5.104) results in $P(w) \sim w^{-\gamma}$, where $\gamma = 2 + \alpha$.

(4) *Growing markets*. One can incorporate the growth feature of the market by introducing explicit time dependence in the distribution of ξ. For instance, the mean $\langle \xi \rangle \equiv a(t)$ may vary in time. The distribution of wealth $P(x, t)$ will then depend explicitly on t. However, in the adiabatic limit, when $a(t)$ varies slowly (such that $a(t-1) \approx a(t)$), we have $P(x, t-1) \approx P(x, t)$. In this limit, thus, $P(x, \tau)$ is identical to the steady-state distribution of the time-independent model, where ξ has an average $\langle \xi \rangle = a(\tau)$.

To demonstrate, let $h(\xi)$ be an exponential distribution with varying average $a(t) = t/T$. In other words, $h(\xi, t) = \exp[-x/a(t)]/a(t)$. From the numerical simulations, calculate the distribution $P(x, T)$ at $t = T$ for different values of T. Since $a(T) = 1$, $P(x, T)$ is compared with the steady-state distribution Eq. (5.102). Numerical investigations suggest that in the quasistatic limit $T \to \infty$ the instantaneous distribution depends only on the instantaneous distribution of ξ.

(5) *Annealed* λ. When the savings propensity of agents changes in time, it is modelled by taking λ as a stochastic variable distributed, say, uniformly in (0, 1). Let $h(\xi) = \exp(-\xi)$. The steady-state distribution of wealth is then $P(x) = \Gamma_2(x) = x\exp(-x)$, which can be proved as follows. If $P(x) = \Gamma_2(x)$, then $P(u = \lambda x) = \exp(-u)$.[1] Thus, the PDF of the right-hand side of Eq. (5.93) is $\Gamma_2(x)$, which is same as the PDF of the left-hand side.

[1] If the PDFs of x and y are $\Gamma_2(x)$ and $\mathcal{U}(y)$, respectively, then the PDF of $u = xy$ is $\exp(-u)$.

This generalized view by Basu and Mohanty (2008) provides a further important insight into the universality of certain features of the basic dynamics of the kinetic exchange models, especially pointing out the cases where conservation is not a necessary criterion. In fact, one can relax the global and even the local conservation rules under specific conditions and still retain the same functional form for the wealth distribution.

5.3.2 Analytic structure of Slanina's model

Equation (4.37) describes a matrix multiplicative stochastic process of vector variable $m(t)$ in discrete time t. Processes of this type are thoroughly studied, e.g. in the context of granular gases. Indeed, if the variables m_i are interpreted as energies corresponding to the i-th granular particle, one can map the process to the mean-field limit of the Maxwell model of inelastic particles. In contrast to assumption $\epsilon > 0$ of the restitution coefficient which models the energy dissipation, now the negative values $\epsilon < 0$ are relevant. This apparently small variation makes a big difference in the analytical treatment of the process, as seen in the treatment of the model in Slanina (2004), which we outline in the following.

The N-particle joint probability distribution $P_N(t; v_1, v_2, \ldots, v_N)$ provides the full information about the process in time t. One can write a kinetic equation using only one- and two-particle distribution functions

$$\begin{aligned} &P_1(t+1;m) - P_1(t;m) \\ &\quad = \frac{2}{N}\left[-P_1(t;m) + \int P_2(t;m_i,m_j)\,\delta((1-\beta+\epsilon)m_i + \beta m_j - m)\,\mathrm{d}m_i\mathrm{d}m_j\right], \end{aligned} \tag{5.105}$$

which can be continued to give eventually an infinite hierarchy of equations of the BBGKY type. A standard approximation is factorizing

$$P_2(t;m_i,m_j) = P_1(t;m_i)P_1(t;m_j), \tag{5.106}$$

breaking the hierarchy on the lowest level by neglecting the correlations between the wealth of the agents, which is induced by the scattering. This approximation, however, becomes exact in $N \to \infty$, which implies that in the thermodynamic limit the one-particle distribution function bears all the necessary information.

One rescales the time as $\tau = 2t/N$ in the thermodynamic limit $N \to \infty$, making the one-particle distribution function $P(\tau;m) = P_1(t,m)$ satisfy a Boltzmann-like kinetic equation

$$\frac{\partial P(m)}{\partial \tau} + P(m) = \int P(m_i)P(m_j)\,\delta((1-\beta+\epsilon)m_i + \beta m_j - m)\,\mathrm{d}m_i \mathrm{d}m_j, \tag{5.107}$$

which describes exactly the process (4.37) in the limit $N \to \infty$. This equation has the same form as the mean-field version for the well-studied Maxwell model of inelastically scattering particles (Ernst and Brito 2002; Krapivsky and Ben-Naim 2002; ben Avraham *et al.* 2003). However, the difference being that here the wealth increases, while in an inelastic gas the energy decreases. This seemingly small difference has deep consequences for the solution of Eq. (5.107). While these dynamical variables correspond rather to energies of the particle, within the framework of the Maxwell model the distributions are expressed in terms of velocities.

The first thing to note is that the average wealth $\bar{m} = \int m\, P(m)\,\mathrm{d}m$ in the process (5.107) grows exponentially and hence Eq. (5.107) has no stationary solution. However, one can look for a quasistationary self-similar solution. A special exactly solvable case $\epsilon = -2\sqrt{\beta} + 2\beta$ provides a hint about possible solutions. Using some standard mathematical tricks, it was shown that the corresponding wealth distribution is of the form $\Phi_1(w) = \frac{1}{\sqrt{2\pi}}\, w^{-5/2} \exp(-\frac{1}{2w})$, which is similar to that obtained in some previous studies (Marsili *et al.* 1998; Bouchaud and Mézard 2000; Solomon and Richmond 2001); see Slanina (2004) for details. In the limit $\epsilon, \beta \to 0$, which indicates infinitesimally small amount of wealth increase and exchange in a single trade event, one can interpret the latter as the limit of continuous trading.

An important feature inferred (Slanina 2004) from the observation was that the system behaves differently for positive and negative ϵ. Indeed, it suggests a singularity at the point of precise conservation of wealth, $\epsilon = 0$.

5.3.2.1 Power law tails

The main concern in empirical studies of wealth distribution is about the shape of tails, which assumes power law forms. Under some specialized assumptions with $\alpha \in (1, 2)$, singularity in some intermediate functions results in the power law tail as $\Phi(w) \sim w^{-\alpha-1}$ for $w \to \infty$, and additionally a transcendental equation for the exponent α, given by

$$(1 + \varepsilon - \beta)^{\alpha} + \beta^{\alpha} - 1 - \varepsilon\alpha = 0. \tag{5.108}$$

While a trivial solution $\alpha = 1$ exists, the power law tail comes from another non-trivial solution, which falls into the desired interval $(1, 2)$ only for certain values of the parameters β and ϵ. There exists an allowed region for the solution. The limits $\epsilon \to 0$, $\beta \to 0$ can be approached, keeping α constant, interpreted as continuous trading, as the amount of wealth exchange and increase in a single trading step is infinitesimally small. From here, one can expect to obtain ordinary differential equations, soluble by standard techniques.

5.3.2.2 Continuous trading limit

Indeed, expanding Eq. (5.108) one obtains the formula relating β and ϵ for fixed α in the continuous trading limit $\beta \to 0, \epsilon \to 0$:

$$\beta = \frac{\alpha - 1}{2} \epsilon^2 + O(\epsilon^3) + O(\epsilon^{2\alpha}) \,. \tag{5.109}$$

The leading correction term to Eq. (5.109) depends on the value of α: for $1 < \alpha < 3/2$ it is of order $O(\epsilon^{2\alpha})$; for $3/2 < \alpha < 2$ it is of order $O(\epsilon^3)$; while in the special point $\alpha = 3/2$ we should include both correction terms, as they are of the same order $O(\epsilon^3)$. See Slanina (2004) for details.

Taking the same limit with fixed α in the non-local case one finally obtains the wealth distribution as

$$\Phi(w) = C\, w^{-\alpha-1} \exp\left(-\frac{\alpha - 1}{w}\right) \tag{5.110}$$

with $C = (\alpha - 1)^\alpha / \Gamma(\alpha)$.

It can be observed that the distribution obtained exhibits the desired power law behaviour for large wealth. Moreover, it has a maximum at a finite value of $w = w_{\max} \equiv (\alpha - 1)/(\alpha + 1)$ and depression for low wealth values. The size of the depletion is determined by the exponential term in Eq. (5.110), i.e. by the same value of α which determines the power in the power law. The idea is that it is the value of the lower bound for the allowed wealth which determines the value of the exponent. Here, however, this result comes purely formally as a result of the analytic computation. In this approach, it is the interplay between wealth increase (taken care of by ϵ) and wealth exchange (taken care of by β) that dictates the value of the exponent α.

5.3.3 Another generalized approach to the kinetic exchange models

While the previous section deals with the essential idea of the physics of dissipative granular gases, it can be extended to further general considerations to include a variety of models under the economic framework of kinetic exchange models (Toscani 2010). Since substantial differences exist between the collision mechanism of classic gas molecules and the trading entities, the interactions can lack the usual microscopic conservation laws for (the equivalents of) energy and momentum. What plays an essential role is the noise in the system. Thus, the rules of 'exchange' are important, but, in contrast to the usual Boltzmann equation, they are defined in an ad hoc manner. As in the classic Boltzmann equation, the relaxation of the system to the Maxwellian equilibrium is shown to be a universal behaviour of the solution, and corresponds to the models as the *macroscopic* statistics of the wealth distribution in the society, to which the solution relaxes. Importantly, while relaxation to equilibrium in the Boltzmann equation is achieved by observing the monotonicity of the entropy functional, the relaxation to the steady-state wealth can be achieved in this case by observing the monotonicity of new convex functionals.

Toscani (2010) points out that there is a remarkable difference between the kinetic exchange models of wealth distribution and the theory of Maxwell molecules (Bobylev 1988): the Maxwell distribution is the universal steady profile for the velocity distribution of molecular gases, while the stationary profiles for wealth can be of various types, and are in general not explicitly known analytically, because they depend heavily on the precise form of the microscopic modelling of trade interactions. Thus, the investigations of the large-time behaviour of the wealth distribution are rather difficult and limited to describe a few analytically accessible quantities. The related body of work done by Giuseppe Toscani and his colleagues mainly considers two types of models – the first type in which the binary trade is microscopically *conservative*, and the second type in which the binary trade is *conservative* in the statistical mean. In both of these situations, the mean wealth in the model Boltzmann equation is preserved, and the formation of a stationary profile is expected.

In the class of point-wise conservative trades, they consider the CC model (Chakraborti and Chakrabarti 2000) and its variants. For the class of conservative in the statistical mean, models with *risky investments* (Cordier *et al.* 2005) are worth mentioning. These analytical techniques easily generalize to a broader class of conservative economic models (Cordier *et al.* 2005; Düring *et al.* 2005, 2008; Pareschi and Toscani 2006; Düring and Toscani 2007; Matthes and Toscani 2008a,b), where trading agents have been treated in the framework of Maxwell-type molecules, and are discussed below.

5.3.3.1 *Boltzmann models for wealth*

To begin with, it is assumed that agents are indistinguishable, as in other models discussed earlier. Then, an agent's 'state' at any instant of time $t \geq 0$ is completely characterized by his current wealth $w \geq 0$. Agents meeting in a trade, their wealths before trading v, w change into v^*, w^* according to the rule

$$v^* = p_1 v + q_1 w, \quad w^* = q_2 v + p_2 w, \tag{5.111}$$

where the *interaction coefficients* p_i and q_i are non-negative random variables. While q_1 denotes the fraction of the second agent's wealth transferred to the first agent, the difference $p_1 - q_2$ is the relative gain (or loss) of wealth of the first agent, attributed to market risks. It is assumed that p_i and q_i have fixed laws, which are independent of v and w, and of time.

In one-dimensional models, the wealth distribution $f(t; w)$ of the ensemble coincides with agent density and satisfies the associated spatially homogeneous Boltzmann equation of Maxwell type

$$\partial_t f + f = Q_+(f, f), \tag{5.112}$$

on the real half-line, $w \geq 0$. The collisional gain operator $Q_+(t; v)$, which quantifies the gain of wealth v at time t owing to binary trades, acts on test functions $\varphi(w)$ as

$$\begin{aligned} Q_+(f, f)[\varphi] :&= \int_{\mathbb{R}_+} \varphi(w) Q_+\big(f, f\big)(w)\, dw \\ &= \frac{1}{2} \int_{\mathbb{R}_+^2} \langle \varphi(v^*) + \varphi(w^*) \rangle f(v) f(w)\, dv\, dw, \end{aligned} \tag{5.113}$$

with $\langle \cdot \rangle$ denoting the expectation with respect to the random coefficients p_i and q_i in Eq. (5.111). The large-time behaviour of the density is heavily dependent on the evolution of the average wealth

$$M(t) := M_1(t) = \int_{\mathbb{R}_+} w f(t; w)\, dw. \tag{5.114}$$

These conservative models are such that the average wealth of the society remains unchanged with time, $M(t) = M$, where the value of M is finite. In terms of the interaction coefficients, this means $\langle p_1 + q_2 \rangle = \langle p_2 + q_1 \rangle = 1$.

The Boltzmann equation (5.112) belongs to the Maxwell type. In the Boltzmann equation for Maxwell molecules the collision frequency is independent of the relative velocity (Bobylev 1988), and the loss term in the collision operator is simply linear, introducing a great simplification which allows the use of most of the well-established techniques developed for the three-dimensional spatially homogeneous Boltzmann equation for Maxwell molecules in the field of wealth redistribution.

Point-wise conservative models Drăgulescu and Yakovenko (2000) and Chakraborti and Chakrabarti (2000) developed the class of strictly conservative exchange models, which preserve the total wealth in each individual trade,

$$v^* + w^* = v + w. \tag{5.115}$$

In the deterministic variant of the CC model (Chakraborti and Chakrabarti 2000), the microscopic interaction is determined by one single parameter $\lambda \in (0, 1)$, which is the global *saving propensity*. Thus, in the interactions, each agent keeps the corresponding λ fraction of his pretrade wealth, while the rest $(1 - \lambda)(v + w)$ is equally shared among the two trade partners,

$$v^* = \lambda v + \frac{1}{2}(1-\lambda)(v+w), \quad w^* = \lambda w + \frac{1}{2}(1-\lambda)(v+w). \tag{5.116}$$

Here, all agents become equally rich eventually (Chatterjee *et al.* 2005c). In the original, non-deterministic variant, the amount $(1-\lambda)(v+w)$ is not equally shared, but in a stochastic way:

$$v^* = \lambda v + \epsilon(1-\lambda)(v+w), \quad w^* = \lambda w + (1-\epsilon)(1-\lambda)(v+w), \tag{5.117}$$

with a random variable $\epsilon \in (0, 1)$.

5.3.3.2 *Conservative in the mean models*

Cordier *et al.* (2005) have introduced the CPT model, which loosens the rule of strict conservation. The idea is that wealth changes hands for a specific reason: one agent intends to *invest* his wealth in some asset, property, etc. in the possession of his trade partner. Typically, such investments bear some risk, and either provide the buyer with some additional wealth or lead to the loss of wealth in a non-deterministic way. An easy realization of this idea (Matthes and Toscani 2008b) couples the previously discussed rules (5.116) with some *risky investment* that yields an immediate gain or loss proportional to the current wealth of the investing agent,

$$v^* = (\lambda + \eta_1)\, v + (1-\lambda)w, \quad w^* = (\lambda + \eta_2)\, w + (1-\lambda)v. \tag{5.118}$$

The coefficients η_1, η_2 are random parameters, which are independent of v and w, and distributed so that always $v^*,\ w^* \geq 0$, i.e. $\eta_1,\ \eta_2 \geq -\lambda$. For centred η_i,

$$\langle v^* + w^* \rangle = (1 + \langle \eta_1 \rangle)v + (1 + \langle \eta_2 \rangle)w = v + w, \tag{5.119}$$

implying conservation of the average wealth. Various specific choices for the η_i can be considered (Matthes and Toscani 2008b). The easiest one leading to interesting results is $\eta_i = \pm\mu$, where each sign comes with probability $1/2$. The factor $\mu \in (0, \lambda)$ should be understood as the *intrinsic risk* of the market: it quantifies the fraction of wealth agents are willing to gamble on.

5.3.3.3 *Boltzmann equilibria*

In conservative markets, where $M(t) = M$, the details of the binary trade determine the profile of the steady-state distribution of wealth. The characteristic function

$$\mathbf{S}(s) = \frac{1}{2}\left(\sum_{i=1}^{2} \langle p_i^s + q_i^s \rangle\right) - 1, \tag{5.120}$$

which is convex in $s > 0$, with $\mathbf{S}(0) = 1$. Also, $\mathbf{S}(1) = 0$ because of the conservation property (5.119). The results from Düring *et al.* (2005) and Matthes and Toscani (2008b) imply that, unless $\mathbf{S}(s) \geq 0$ for all $s > 0$, any solution $f(t; w)$ tends to a steady wealth distribution $P_\infty(w) = f_\infty(w)$, which depends on the initial wealth distribution only through the conserved mean wealth $M > 0$. Moreover, exactly one of the following is true:

(A) if $\mathbf{S}(\alpha) = 0$ for some $\alpha > 1$, then $P_\infty(w)$ has a *Pareto tail* of index α;
(B) if $\mathbf{S}(s) < 0$ for all $s > 1$, then $P_\infty(w)$ has a *slim tail*;
(C) if $\mathbf{S}(\alpha) = 0$ for some $0 < \alpha < 1$, then $P_\infty(w) = \delta_0(w)$, a *Dirac delta* at $w = 0$.

In case (A), exactly the moments $M_s(t)$ with $s > \alpha$ blow up as $t \to \infty$, giving rise to a Pareto tail of index α. It has to be emphasized that $f(t; w)$ possesses finite moments of all orders at any finite time. The Pareto tail forms in the limit $t \to \infty$.

In case (B), all moments converge to limits $M_s(t) \to M_s^*$, so the tail is slim.

In case (C), all moments $M_s(t)$ with $s > 1$ blow up. The underlying process is a separation of wealth as time increases: while more and more agents become extremely poor, fewer and fewer agents possess essentially the entire wealth of the society.

One study used Fourier transformations extensively to derive these results, and this works very effectively to treat collision kernels of Maxwellian type (Bobylev 1988). Particularly, the Fourier representation is specially adapted to the use of various Fourier metrics.

According to the collision rule (5.111), the transformed gain term reads

$$\widehat{Q}_+\left(\widehat{f}, \widehat{f}\right)(\xi) = \frac{1}{2}\left\langle \widehat{f}(p_1\xi)\widehat{f}(q_1\xi) + \widehat{f}(p_2\xi)\widehat{f}(q_2\xi)\right\rangle. \tag{5.121}$$

The initial conditions turn into

$$\widehat{f_0}(0) = 1 \quad \text{and} \quad \widehat{f_0}'(0) = iM.$$

Hence, the Boltzmann equation (5.112) can be rewritten as

$$\frac{\partial \widehat{f}(t;\xi)}{\partial t} + \widehat{f}(t;\xi) = \frac{1}{2}\left\langle \widehat{f}(p_1\xi)\widehat{f}(q_1\xi) + \widehat{f}(p_2\xi)\widehat{f}(q_2\xi)\right\rangle. \tag{5.122}$$

Equation (5.122) can be easily treated from a mathematical point of view owing to the well-known techniques introduced so far to study the Boltzmann equation for Maxwell molecules and its caricatures, mainly the Kac equation (Kac 1959).

In this chapter, we have dealt with the analytical studies of the kinetic exchange models. In particular, we have shown that the entropy maximization process gives rise to the inequality of the quantity exchanged. In the following chapter, we will give the microeconomic formulation of such models.

6

Microeconomic foundation of the kinetic exchange models

In the earlier chapters, we introduced kinetic exchange models and discussed the mathematics behind them. However, one major point that is missing in the earlier discussions is the choice behaviour of the agents. The outcomes of the stochastic process do not reflect any optimization mechanism on the part of the agents. In this chapter, we will provide a simple economic model which intends to capture the basic features of the kinetic exchange models. We start with some usual assumptions regarding the preference pattern of the agents and the market mechanism. Eventually, it will be shown that the outcomes are exactly the same as those obtained in the kinetic exchange models, thus providing an elementary (but non-unique) way of interpreting the stochastic money evolution equations in economic terms. After that, we will discuss the dynamic features of the asset distribution in the economy if it has time-varying macroeconomic variables. To be precise, we will discuss a possibility of inequality reversal (as has been postulated and discussed in Kuznets (1955, 1965) and Angle (2010)) in the same framework.

6.1 Derivation of the basic kinetic exchange model

Following Chakrabarti and Chakrabarti (2009), we consider an N-agent exchange economy in discrete time. At every point of time, exactly two agents are randomly chosen from the pool of N agents, i.e. each agent has equal probability of being chosen for trade. The exact trading mechanism is described below. After they trade, the agents part and leave the market. In the next period again two agents are chosen for trade and the same process is repeated until the distribution of their assets reaches a steady state. More specifically, at each point in time, each of the chosen agents produces a single perishable commodity. These commodities are different from each other, and therefore the agents will have an incentive to trade with each other. We assume that money exists in this economy to facilitate transactions. These agents care for their future consumptions and hence they care about their savings in the current period as well. Each of these agents is endowed with an initial amount of money (the only type of non-perishable asset considered), which is assumed to be unity for every agent for simplicity. The initial amount of money could be varied of course. But that does not change the qualitative behaviour of the model. At each time step, two agents meet randomly to carry out transactions according to the utility maximization principle. The agents are endowed with a time-dependent preference structure, i.e. the parameters of the utility function can vary over time (Silver *et al.* 2002).

Let us discuss what happens at any point of time t. Suppose two agents from the pool of N agents have been chosen randomly to trade among themselves. For notational simplicity we denote them as agents 1 and 2. Assume that agent 1 produces Q_1 amount of commodity 1 only and agent 2 produces Q_2 amount of commodity 2 only and the amounts of money they possess at time t are $m_1(t)$ and $m_2(t)$, respectively (clearly, $m_i(1) = 1$ for all i). Both of them will be willing to trade and buy the other's good by selling a fraction of their own productions and also with the money that they have. In general, at each time step there would be a net transfer of money from one agent to the other. This is where our final interest lies. Below, we define the trading process. To keep notations simple, we get rid of the time index t and define $m_i(t) = M_i$ and $m_i(t+1) = m_i$ in this subsection. The utility functions are defined (without the time indices) as follows: for agents 1 and 2, respectively,

$$U_1(x_1, x_2, m_1) = x_1^{\alpha_1} x_2^{\alpha_2} m_1^{\lambda} \quad \text{and} \quad U_2(y_1, y_2, m_2) = y_1^{\beta_1} y_2^{\beta_2} m_2^{\lambda}, \tag{6.1}$$

where the arguments in both of the utility functions are consumption of the first (i.e. x_1 and y_1) and second (i.e. x_2 and y_2) good and the amount of money they possess, respectively. See Mas-Colell *et al.* (1995) for an excellent introduction to the theory of choices and the usage of utility functions to represent them. For simplicity, we assume that the utility functions are of the Cobb–Douglas form with the sum of the powers normalized to 1, i.e.

$$\alpha_1 + \alpha_2 + \lambda = 1 \quad \text{and} \quad \beta_1 + \beta_2 + \lambda = 1. \tag{6.2}$$

Clearly, we are interpreting λ as the power of money in the utility function. Later on we will find that the same term will stand for the savings propensity in the money evolution equations. We have already described the objective functions which are to be maximized. But, of course, these are not unbounded optimization exercises. The agents' consumption decisions are bounded by their respective purchasing powers. Let p_1 and p_2 be the commodity prices to be determined in the market. The budget constraints are defined as follows: for agents 1 and 2, respectively, the budget constraints are

$$p_1x_1 + p_2x_2 + m_1 \leq M_1 + p_1Q_1 \quad \text{and} \quad p_1y_1 + p_2y_2 + m_2 \leq M_2 + p_2Q_2. \tag{6.3}$$

This means that the amount that agent 1 can spend for consuming x_1 and x_2 added to the amount of money that he holds after trading at time $(t+1)$ (i.e. m_1) cannot exceed the amount of money that he has at time t (i.e. M_1) added to what he earns by selling the good he produces (i.e. Q_1), and the same is true for agent 2. Note that money acts as the *numéraire* good as usual and, hence, its price is 1. Subject to their respective budget constraints, the agents maximize their respective utility functions and the pricing mechanism (*invisible hand*) works to clear the market for both goods (i.e. total demand equals total supply for both goods at the equilibrium prices), i.e. agent 1's problem is to maximize his utility $U_1(x_1, x_2, m_1)$ subject to $p_1.x_1 + p_2.x_2 + m_1 = M_1 + p_1.Q_1$, and for agent 2, to maximize $U_2(y_1, y_2, m_2)$ subject to $p_1.y_1 + p_2.y_2 + m_2 = M_2 + p_2.Q_2$. Let us solve the problem for the first agent using the Lagrange multiplier technique,

$$\mathcal{L}_1 = x_1^{\alpha_1}x_2^{\alpha_2}m_1^{\lambda} - \mu_1(p_1x_1 + p_2x_2 + m_1 - M_1 - p_1Q_1).$$

Equating the first derivatives (with respect to x_1, x_2, m_1 and the Lagrange multiplier μ_1) with zero, one can derive the demand functions of the first agent as the following:

$$x_1^* = \alpha_1\frac{(M_1 + p_1Q_1)}{p_1}, \quad x_2^* = \alpha_2\frac{(M_1 + p_1Q_1)}{p_2},$$
$$m_1^* = \lambda(M_1 + p_1Q_1).$$

By solving the problem for the second agent via

$$\mathcal{L}_2 = y_1^{\beta_1}y_2^{\beta_2}m_2^{\lambda} - \mu_2(p_1y_1 + p_2y_2 + m_2 - M_2 - p_2Q_2),$$

we derive the second agent's demand functions, which are

$$y_1^* = \beta_1\frac{(M_2 + p_2Q_2)}{p_1}, \quad y_2^* = \beta_2\frac{(M_2 + p_2Q_2)}{p_2},$$
$$m_2^* = \lambda(M_2 + p_2Q_2).$$

The market clearing conditions are $x_1^* + y_1^* = Q_1$ and $x_2^* + y_2^* = Q_2$ (i.e. demand matches supply in both the markets at equilibrium prices). One noteworthy feature is that it was not necessary to choose these two particular markets. We could have chosen any two markets to clear, e.g. we could have chosen the money market and the market for commodity 1 to clear. However, *Walras' law* says that if all but one market clears then the rest also has to be cleared at the same price vector, i.e. we do not worry about the third market. It is automatically cleared (see point (1) below). By substituting the values of x_1^*, x_2^*, y_1^* and y_2^* and by solving these two equations we get market clearing prices $(\hat{p}_1, \hat{p}_2)$, where

$$\hat{p}_1 = \frac{(\lambda\alpha_1 + \beta_1(1-\lambda))\,M_1 + \beta_1 M_2}{\lambda Q_1(1-\alpha_1+\beta_1)}$$

and

$$\hat{p}_2 = \frac{\alpha_2 M_1 + ((1-\lambda)\alpha_2 + \lambda\beta_2)\,M_2}{\lambda Q_1(1-\alpha_1+\beta_1)}.$$

By substituting $(\hat{p}_1, \hat{p}_2)$ in the money demand equations, we get

$$\left.\begin{aligned} m_1^* &= \lambda M_1 + \frac{\lambda\alpha_1 + (1-\lambda)\beta_1}{1-\alpha_1+\beta_1} M_1 + \frac{\beta_1}{1-\alpha_1+\beta_1} M_2, \\ m_2^* &= \lambda M_2 + \frac{\alpha_2}{1-\alpha_1+\beta_1} M_1 + \frac{\lambda\beta_2 + (1-\lambda)\alpha_2}{1-\alpha_1+\beta_1} M_2. \end{aligned}\right\} \tag{6.4}$$

Now, we denote m_i^* as $m_i(t+1)$ and M_i as $m_i(t)$ (for $i = 1, 2$). The above set of equations can be rewritten as

$$\left.\begin{aligned} m_1(t+1) &= \lambda m_1(t) + \theta_{11} m_1(t) + \theta_{12} m_2(t), \\ m_2(t+1) &= \lambda m_2(t) + \theta_{21} m_1(t) + \theta_{22} m_2(t), \end{aligned}\right\} \tag{6.5}$$

by appropriately defining θ_{ij}'s (for i, j=1, 2). We are now in a position to discuss the outcomes of such a trading process.

(1) At optimal prices $(\hat{p}_1, \hat{p}_2)$, $m_1(t) + m_2(t) = m_1(t+1) + m_2(t+1)$, i.e. demand matches supply in all markets at the market-determined prices in equilibrium. Since money is also treated as a commodity in this framework, its demand (i.e. the total amount of money held by the two persons after trade) must equal what was supplied (i.e. the total amount of money held by them before trade). Recall that we did not allow money to be created or destroyed in this economy just as is the case in the 'conservative' kinetic exchange market models.

(2) We now make a very restrictive assumption that $\alpha_i = \beta_i$ for $i = 1, 2$. This assumption drastically simplifies the money evolution equations:

$$\left.\begin{aligned} m_1(t+1) &= \lambda m_1(t) + \alpha_1(m_1(t) + m_2(t)), \\ m_2(t+1) &= \lambda m_2(t) + \alpha_2(m_1(t) + m_2(t)). \end{aligned}\right\} \tag{6.6}$$

Next, to introduce randomness in the choice behaviour we assume that α_1 in the utility function can vary randomly over time with λ remaining constant. This in turn implies that α_2 also varies randomly over time with the restriction that the sum of α_1 and α_2 is a constant $(1-\lambda)$ because of the normalization

$\alpha_1 + \alpha_2 + \lambda = 1$. Now, in the money demand equations derived above, if $\alpha_1/(\alpha_1 + \alpha_2)$ is substituted by ϵ, the money evolution equations become:

$$\left.\begin{aligned} m_1(t+1) &= \lambda m_1(t) + \epsilon(1-\lambda)[m_1(t) + m_2(t)], \\ m_2(t+1) &= \lambda m_2(t) + (1-\epsilon)(1-\lambda)[m_1(t) + m_2(t)]. \end{aligned}\right\} \tag{6.7}$$

This is *exactly* the CC model (Chakraborti and Chakrabarti 2000) with λ (i.e. the power of money in the utility function) as the savings propensity. For a fixed value of λ, if α_1 (or α_2) is a random variable with uniform distribution over the domain $[0, 1-\lambda]$, then ϵ is also uniformly distributed over the domain $[0, 1]$.

(3) For the limiting value of λ in the utility function (i.e. $\lambda \to 0$), the money evolution equation describing the random sharing of money without savings is retrieved, as used in the model of Drăgulescu and Yakovenko (2000).

(4) If $\alpha_i \neq \beta_i$ for $i = 1, 2$, then evidently the θ_{ij} terms in equation 6.5 are correlated, which gives rise to a new possibility of having a generalization of the earlier framework in the direction of having correlated returns. We will have more to say on this point later.

(5) An important property of the above-mentioned model is that there are non-trivial interactions among the agents and hence the asset evolution equations are coupled. This is the crucial aspect in which the current model is different from a previous known work that has a similar line of thinking (Silver *et al.* 2002) in which the asset evolution equation for the i-th agent depends on its own assets.

Let us now discuss the moments generated by the above-mentioned stochastic difference equations as they will give us a qualitative picture of the distributions of money.

6.1.1 Random exchange

For completely random sharing of assets, there is no savings, i.e. $\lambda = 0$. Hence, the exchange equations look like:

$$\left.\begin{aligned} m_i(t+1) &= \epsilon(m_i(t) + m_j(t)), \\ m_i(t+1) &= (1-\epsilon)(m_i(t) + m_j(t)), \end{aligned}\right\} \tag{6.8}$$

where $\epsilon \in [0, 1]$ and uniform (one can also consider $\epsilon \in [\delta, 1-\delta]$ with $0 \leq \delta < 0.5$, and the qualitative features still hold good). Evidently $\langle m \rangle = 1$ since there is no creation or destruction of money and we assumed that all agents (identical in all aspects) start with exactly one unit of money. Also, in the steady state $\Delta m_i = \Delta[\epsilon(m_i + m_j)] = \langle x^2 \rangle - \langle x \rangle^2$, where $x = [\epsilon(m_i + m_j)]$. Note that

$\langle x \rangle = 1$. Hence,

$$\Delta m = \langle \epsilon^2 \rangle \langle m_i^2 + m_j^2 + 2m_i m_j \rangle - 1. \tag{6.9}$$

Using the fact that m_i and m_j are uncorrelated (in a very large system, i.e. where $N \to \infty$) and $\Delta\epsilon = \langle \epsilon^2 \rangle - 1/4$, we get

$$\Delta m = \left(\Delta\epsilon + \frac{1}{4} \right) (2\Delta m + 4) - 1. \tag{6.10}$$

Simplifying, we get $\Delta m = \frac{4\Delta\epsilon}{\frac{1}{2} - 2\Delta\epsilon}$.

6.1.2 Exchange with savings

Here also, $\langle m \rangle$ =1. In this context, the variance of the distribution will be given by the following equation (apply variance operator on both sides of equation 6.7):

$$\Delta m = \lambda^2(\Delta m + 1) + 2(1 - \lambda)^2 \left(\Delta\epsilon + \frac{1}{4} \right) (\Delta m + 2) + \lambda(1 - \lambda)(\Delta m + 2) - 1.$$

Here, Δ stands for the second central moment (variance). Since $\epsilon \in [0, 1]$ and uniformly distributed, $\Delta\epsilon = 1/12$. Thus one gets

$$\Delta m = \frac{(1 - \lambda)^2}{(1 - \lambda)(1 + 2\lambda)}. \tag{6.11}$$

Hence, if $\lambda \neq 1$, $\Delta m = (1 - \lambda)/(1 + 2\lambda)$, as in Patriarca *et al.* (2004). Thus, $\Delta m = 1$ for $\lambda = 0$, which is the case for an exponential distribution, and for $0 \leq \lambda < 1$ the distribution is well approximated by

$$p(m) = \frac{m^{\kappa_1 - 1} e^{-\kappa_2 m}}{\Gamma(\kappa_1) \kappa_2^{-\kappa_1}},$$

with $\kappa_1 = (1 + 2\lambda)/(1 - \lambda)$ and $\kappa_2 = \kappa_1$ as is conjectured in Patriarca *et al.* (2004) (see Section 5.1.1). For $\lambda \to 1$, by applying l'Hôpital's rule one gets $\Delta m = 0$, explaining why the steady-state distribution tends to a delta function as the rate of savings, i.e. $\lambda \to 1$ as widely observed in simulations (Chakraborti and Chakrabarti 2000; Patriarca *et al.* 2004). So far, while deriving the above model, we have assumed that the agents are producing only one commodity at every time point. However, we can present the same model by incorporating the idea of risk aversion explicitly when each agent produces a vector of commodities. Below we discuss the notion of correlation in the returns from trading of several commodities simultaneously.

6.2 Production of a vector of commodities

We begin with a simple calculation. Suppose an agent invests a certain amount of money in K number of assets in which the returns are stochastic. More precisely, let us assume that the returns (ϵ_k) are independent and identically distributed variables with finite mean (μ) and variance (σ^2). The problem for the agent is to decide what fractions (f_k) of his money holding he would invest in each asset k for $k = 1, 2, \ldots, K$. Assuming risk aversion, the problem is to minimize the variance of his portfolio ($\sum_k \epsilon_k f_k$) or, formally, the problem is to minimize

$$\sigma^2(\epsilon)\left(f_1^2 + f_2^2 + \cdots + f_K^2\right)$$

subject to the condition that

$$f_1 + f_2 + \cdots + f_K = 1.$$

Clearly, the solution would be $f_k^* = 1/K$ for all k. Now, consider the following set of stochastic difference equations representing how the money holding changes among the (without savings) agents over time:

$$\left.\begin{aligned} m_i(t+1) &= \epsilon[m_i(t) + m_j(t)], \\ m_j(t+1) &= (1-\epsilon)[m_i(t) + m_j(t)]. \end{aligned}\right\} \tag{6.12}$$

While deriving this set of equations, it had been assumed that each of the agents produced a single non-storable commodity and money acted as an asset that helps to make transactions. However, we can generalize the situation by assuming that each of the agents produces a vector of commodities and engages in trading with the other, then it is perfectly possible for a risk-averse agent to diversify his money holding at time t following the above calculation, instead of putting all his money in trading of a single commodity. The market has the following structure. Each agent produces K ($K \geq 1$) number of commodities and each of these commodities is different from the other. Hence, the agents would be willing to trade with each other. Above in Section 6.1, we have dealt with the case where $K = 1$, i.e. each agent produces a single commodity and it shows that Eq. 6.12 captures the basic process of money exchange in such an economy with zero savings propensity. Here, we consider the case where $K \geq 1$. Clearly, the risk-averse agents would diversify their money holding in order to minimize the risk from trading. The mode of trading is such that, at each instant, two randomly chosen agents engage in trading, each producing K number of different goods so that, in total, $2K$ number of goods are traded at each instant. For trading the k-th pair of goods ($k = 1, 2, \ldots, K$), the i-th and the j-th agent use m_i/K and m_j/K amounts of money, respectively, because we have already shown that the variance (risk) minimizing choice is to diversify equally among all assets. So the money evolution equations become the generalization of Eq. 6.12, namely

$$\left.\begin{aligned} m_i(t+1) &= \frac{\sum_k \epsilon_k}{K}[m_i(t)+m_j(t)], \\ m_j(t+1) &= \left(1-\frac{\sum_k \epsilon_k}{K}\right)[m_i(t)+m_j(t)], \end{aligned}\right\} \tag{6.13}$$

for all possible integer values of K. If K is 1, then we get back Eq. 6.12, which implies that the steady-state distribution of money would be exponential. In the other extreme for $lim \quad K \to \infty$, by applying the Lindeberg–Levy central limit theorem, we have

$$\left(\frac{\Sigma_k \epsilon_k}{\sqrt{K}}\right) \sim N(\mu, \sigma^2),$$

where μ and σ^2 are finite for uniformly distributed variables ϵ_k. This in turn implies that

$$\frac{\Sigma_k \epsilon_k}{K} \sim N\left(\mu, \frac{\sigma^2}{K}\right).$$

Hence, the distribution is a Δ function at μ for large K. The resulting distribution of money would also be a Δ function, i.e. perfect equality will be achieved. For finite values of K greater than unity, the distribution of money would have positive skewness (see Chakrabarti and Chakrabarti 2010a). It may be noted that the assumption of the returns having the same mean and variance along with independence among themselves is not very realistic. Further generalizations are possible. However, our aim was to show that, even if we do not consider savings propensity, it is possible to generate positively skewed distributions in money holding from the very basic random exchange equations (Eq. 6.8) if we consider multisectoral trade. But, of course, the shifts in the distributions are discrete (for $K = 1, 2, 3, \ldots$) and not continuous as in the CC model.

6.3 A generalized version of the CC model

Recall that we derived a set of stochastic money evolution equations as (Eq. 6.5)

$$\left.\begin{aligned} m_1(t+1) &= \lambda m_1(t) + \theta_{11} m_1(t) + \theta_{12} m_2(t), \\ m_2(t+1) &= \lambda m_2(t) + \theta_{21} m_1(t) + \theta_{22} m_2(t). \end{aligned}\right\} \tag{6.14}$$

The presence of a positive savings propensity is evident in the above equation. Assuming that α_i and β_i (for $i = 1, 2$) are random variables (because of the time dependence of preference), we see that the θ_{ij}'s are correlated (for $i, j = 1, 2$). Hence, the money evolution equations consists of two correlated random terms. We intend to use this model to shed light on the dynamic behaviour of the steady-state distribution of money holding in an economy which has time-dependent savings behaviour and riskiness. For simplicity and tractability, we assume the θ_{ij}'s to be correlated in the following form:

$$\begin{aligned} m_i(t+1) &= \lambda m_i(t) + \omega_1(1-\lambda)m_i(t) \\ &\quad + [\alpha\omega_1 + (1-\alpha)\omega_2]\,(1-\lambda)m_j(t), \\ m_j(t+1) &= \lambda m_j(t) + (1-\omega_1)(1-\lambda)m_i(t) \\ &\quad + [1-\alpha\omega_1 - (1-\alpha)\omega_2]\,(1-\lambda)m_j(t), \end{aligned}$$

where $\omega_1, \omega_2 \sim \text{uniform}[0, 1]$ and independent. That is, instead of having non-linear terms involving two random variables, we consider linear combinations of them while keeping the rule of conservation of money intact. We closely follow the formulation given in Chakrabarti and Chakrabarti (2010a), which studied this model in detail. Note that in the above set of equations ω_1 and $(\alpha\omega_1 + (1-\alpha)\omega_2)$ are the two correlated random terms. Later, we will see that the distributional assumptions of ω_1 and ω_2 will help us to calculate the moments of the resultant steady-state distributions of money very easily. The savings propensity and the degree of correlation between the stochastic terms are denoted by λ and α, respectively, and both can vary between 0 and 1, leading to different steady-state distributions as shown in Fig. 6.1. Technically, α is not the *correlation coefficient*. It is a parameter that will be helpful to tune the correlation between the two random terms. Several points are to be noted.

(1) If $\lambda = 0$ and $\alpha = 1$, then we have the very basic framework of an ideal gas, which gives rise to a purely exponential distribution (Gibbs distribution: $p(m) \sim e^{-m/T}$ with $T = 1$ in this case); see Yakovenko and Barkley Rosser (2009) and Drăgulescu and Yakovenko (2000).
(2) If $\lambda = 0$ and $\alpha = 0$, then we have a model with two uncorrelated stochastic terms. This model has been studied and solved in Majumdar *et al.* (2000). This model gives rise to a probability distribution characterized by a gamma probability density function of the form $p(m) \sim 4me^{-2m}$.
(3) If $lim\ \lambda \to 1$, then the distribution would be a delta function.
(4) If only $\alpha = 1$, the above model reduces to the so-called CC model (Chakraborti and Chakrabarti 2000), which gives rise to gamma function-like behaviour.
(5) If only $\alpha = 0$, then we have a new model which has savings propensity (CC model) and two uncorrelated random terms (see point (2) above).

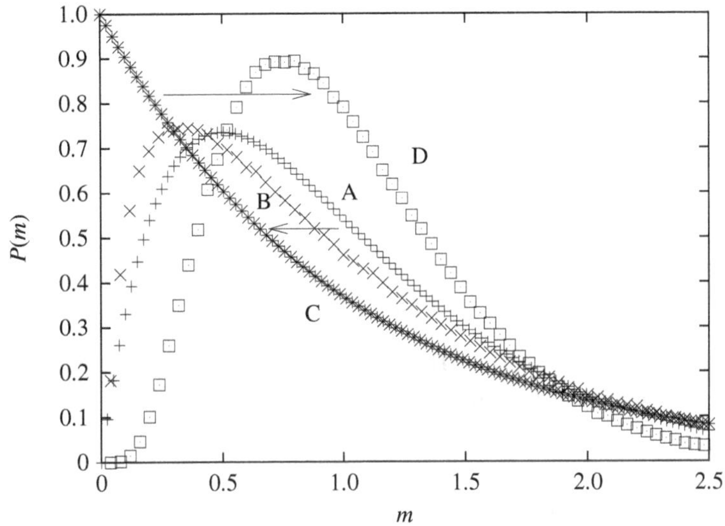

Figure 6.1 Steady-state distributions of money for different values of the parameters. A, $\lambda = 0$ and $\alpha = 0$. B, $\lambda = 0$ and $\alpha = 0.7$. C, $\lambda = 0$ and $\alpha = 1$. D, $\lambda = 0.5$ and $\alpha = 1$. As the correlation goes up the distribution becomes more skewed to the left (from A to B to C; see the arrow). Then as the savings propensity goes up, it moves in the opposite direction (from C to D; see the arrow). All simulations are done for $O(10^6)$ time steps with 100 agents and averaged over $O(10^3)$ time steps. Reproduced from Chakrabarti and Chakrabarti (2010a).

6.4 Inequality reversal

We can measure inequality by a number in indices. See the next chapter for a short discussion on this topic. However, the most useful one (and the easiest one to implement as will be seen shortly) in this case is simply the coefficient of variation, which is basically the standard deviation of the distribution normalized by the mathematical expectation. The economy is modelled in such a way that the expectation is always set equal to unity (recall that all agents are initially endowed with unit amount of money and the economy under study is a conserved one, i.e. money is neither created nor annihilated in this economy). We can calculate the moments recursively as we did in the earlier section (see Repetowicz *et al.* (2005) for more on finding the moments). We consider the i-th agent's money evolution equation only

$$m_i(t+1) = [\lambda + \omega_1(1-\lambda)]m_i(t) + [\alpha\omega_1 + (1-\alpha)\omega_2](1-\lambda)m_j(t). \quad (6.15)$$

It is easy to show that $\langle m \rangle = 1$. Formally, by taking expectation over both sides of Eq. 6.15, we get

$$\langle m_i(t+1)\rangle = \left[\lambda + \frac{1}{2}(1-\lambda)\right]\langle m_i(t)\rangle + \frac{1}{2}(1-\lambda)\langle m_j(t)\rangle.$$

Note that $\langle m_j(t)\rangle = \sum_{j=1}^{N} m_j(t)/N = 1$. Since the expected money holding is free of the time index, the result readily follows. The next result is

$$\Delta m = \frac{2(1-\lambda)\left[\frac{\alpha(1-\lambda)}{3} + \frac{\lambda}{2} + \frac{(1-\lambda)(1-\alpha)}{4}\right]}{1-z} - 1,$$

where $z = (1-\lambda)^2\left(\frac{1}{3} + \frac{\lambda}{(1-\lambda)^2} + \frac{\alpha^2+(1-\alpha)^2}{3} + \frac{\alpha(1-\alpha)}{2}\right)$. The proof involves a little bit of algebra. It follows from the definition of variance that

$$\Delta m_i = \langle m_i^2\rangle - (\langle m_i\rangle)^2,$$

where m_i is given by Eq. 6.15. By substituting m_i in the left-hand side of the expression of variance and noting that $\langle m\rangle = 1$, we get

$$\Delta m_i(t+1) = \langle[(\lambda + \omega_1(1-\lambda))m_i(t) + (\alpha\omega_1 + (1-\alpha)\omega_2)(1-\lambda)m_j(t)]^2\rangle - 1.$$

Here, we use the fact that, in the steady state, the variance of the distribution should be free of the time and the agent indices. Also, since $\omega_i \sim$ uniform [0,1], $\langle\omega_i\rangle = 1/2$ and $\Delta\omega_i = 1/12$ (for i =1, 2). On simplification, we get

$$\Delta m = z(\Delta m + 1) + 2(1-\lambda)\langle(\lambda + \omega_1(1-\lambda))\,(\alpha\omega_1 + (1-\alpha)\omega_2)\rangle - 1,$$

where $z = (1-\lambda)^2\left(\frac{1}{3} + \frac{\lambda}{(1-\lambda)^2} + \frac{\alpha^2+(1-\alpha)^2}{3} + \frac{\alpha(1-\alpha)}{2}\right)$. On further simplification, we get

$$\Delta m = \frac{2(1-\lambda)\left[\frac{\alpha(1-\lambda)}{3} + \frac{\lambda}{2} + \frac{(1-\lambda)(1-\alpha)}{4}\right]}{1-z} - 1, \tag{6.16}$$

where z is defined as above. Clearly the variance is a function of λ and α only. Now, we make use of two observations. First, for a sustainable growth the savings propensity cannot be too low (see Barrow and Sala-i Martin (2004) for more on this topic). If an economy starts from a subsistence-level consumption (where the whole income is consumed with nothing left as savings), it has to increase the savings propensity over time to achieve prosperity. The second observation is that the modern markets are characterized by correlated returns with fluctuations (Sornette 2004) in the most efficient state (Bak 1996). The implication is that both λ and α may increase over time unidirectionally. By plugging different values of λ and α in the expression of variance, one can find how the inequality index changes over time with increases in the parameters. Note that, since the parameters are ranging between 0 and 1, the parameter space is a square with unit length (Fig. 6.2 shows the relevant region). For simplicity consider the following scenario. Assume that the economy starts from a situation where there is no savings and also the returns are completely uncorrelated. Over time the savings propensity increases. So does the correlation in the asset returns. At the end the economy has very high savings

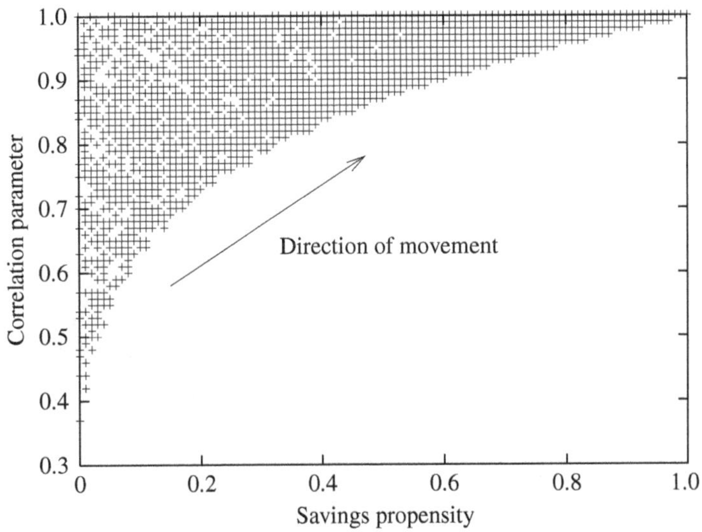

Figure 6.2 The (λ, α) parameter space. If the economy moves through the shaded region (the region above the curve $\alpha = \lambda^{1/5}$) then it shows inequality reversal. Reproduced from Chakrabarti and Chakrabarti (2010a).

propensity and very high correlation. In short, the path followed by the economy starts from the origin ($\lambda = 0$, $\alpha = 0$) and ends at ($\lambda = 1$, $\alpha = 1$). The simplest functional relationship between λ and α satisfying the above assumption is of the form

$$\alpha = \lambda^{\frac{1}{\tau}}, \tag{6.17}$$

where τ is a positive number. It is numerically seen that, for $\tau \geq 5$, the economy shows a very prominent inequality reversal (Fig. 6.3). It should be noted, however, that if the economy follows some other paths in the parameter space, then it may show other types of behaviour as well. For the sake of completeness, we also provide Monte Carlo simulation results of the Kuznets curves, in which the inequality is measured in terms of the Gini concentration ratio. The definition of the measure G is the following (Kleiber and Kotz 2003),

$$G \equiv \frac{\sum_{i=1}^{N} \sum_{j=1}^{N} \mid m_i - m_j \mid}{2\mu N(N-1)}, \tag{6.18}$$

where N is the number of agents (which is set to 100), μ is the money per agent (which is set to unity) and m_i is the money holding of the i-th agent. Figure 6.4 shows the rise and the subsequent fall in the Gini concentration ratio. In the original formulation, the Kuznets curve depicted the changes of income distribution with the changes in per capita income. However, that possibility is lost in the current

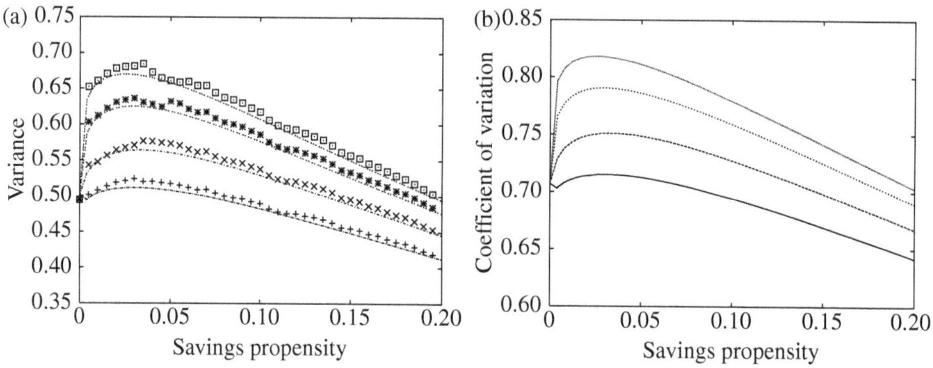

Figure 6.3 (a) The changes in the variance of the distribution (Δm) are shown with changes in the savings propensity (λ) (from below, $\tau = 5, 7, 10$ and 13). The Monte Carlo simulation results agree with the theoretical curves (dotted lines) obtained from Eqs. 6.16 and 6.17. (b) The Kuznets curve in terms of the coefficient of variation, i.e. $\sqrt{\Delta m}$ (from below, $\tau = 5, 7, 10$ and 13): inequality increases and then falls. Reproduced from Chakrabarti and Chakrabarti (2010a).

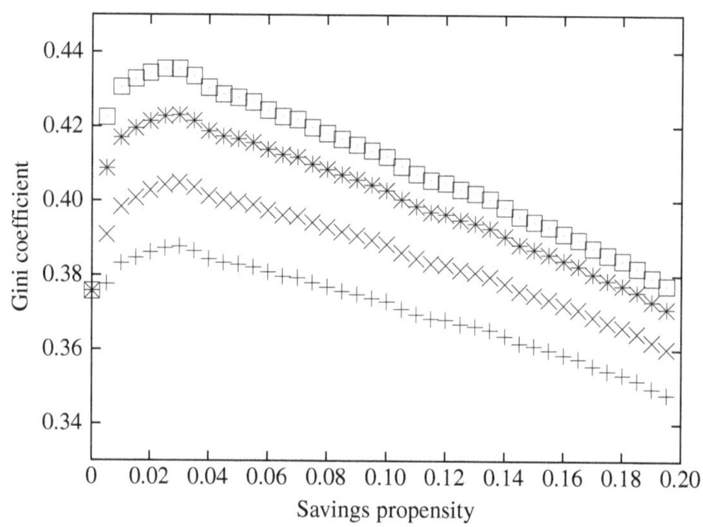

Figure 6.4 The Kuznets curve in terms of the Gini concentration ratio (from below, $\tau = 5, 7, 10$ and 13): inequality increases and then falls. All Monte Carlo simulations are done for $O(10^5)$ time steps with 100 agents and averaged over $O(10^2)$ time steps. Reproduced from Chakrabarti and Chakrabarti (2010a).

framework because the trading process that we have considered is a conservative one, implying that the average money holding in this model remains fixed over time. There is no growth in the economy. Hence, instead of average income, we consider the changes in the savings propensity and the correlation among the markets and we trace the corresponding effects on the inequality in money holding. In such cases, it is clearly seen that the economy shows Kuznets-type, i.e. inverted U-shaped, behaviour. From what we have shown above, it is clear that the kinetic exchange models can accommodate the idea of time-varying inequality very easily. Almost all the economies show differences in inequalities over time. However, the validity of the claim of an exact inverted U-shaped inequality path is not beyond doubt.

6.5 Global market

Another interesting possibility would be to consider a global market. The main problem with the earlier formulation is that we cannot derive an explicit formula to describe the steady-state probability distributions generated by Eq. 6.7 except at $\lambda = 0$. Recall that, the way we had formulated, the price mechanism could work only locally (matching the bipartite supply and demand) and the markets were also cleared locally. Now we focus on global market clearance via the price mechanism. It will be shown that the money evolution equations would be non-coupled and hence solvable under mild distributional conditions. Below we consider a highly simplified model to capture the idea with a minimal number of parameters. We assume that, subject to the individual budget constraints, each agent maximizes its utility and allocates income accordingly between present and future consumptions. Both the present and future consumptions are represented by the amount of money spent on the present and future consumptions. We ignore any discount factor between current and future consumptions. On the production side, they can invest in production and get their returns accordingly. Formally there are N agents in the economy each taking part in production and consumption. A typical agent's behaviour at any time step t is analysed as follows:

(1) Each agent has to maximize utility subject to his budget constraint. For simplicity, the utility function is assumed to be of Cobb–Douglas type. Briefly, at time t the i-th agent's problem is to maximize $u(f, c) = f^{\lambda} c^{(1-\lambda)}$ subject to $f + c = m(t)$, where f is the amount of money kept for future consumption, c is the amount of money to be used for current consumption and $m(t)$ is the amount of money holding at time t. This is a standard utility maximization problem and, solving it by Lagrange multiplier, one gets the optimal allocation as $c^* = (1 - \lambda)m(t)$ and $f^* = \lambda m(t)$. Note the difference between the utility functions considered in Section 6.1 and the one we are considering now. Here, we assume that the agents derive utility from holding assets, which is not really a particularly compelling assumption. But it simplifies the model a lot.

2 The i-th agent invests $(1 - \lambda_i)m_i(t)$ in the market and produces an output vector $y_i(t)$, which he sells in the market at market-determined price vector p_t, being the same for everybody. Here we ignore the exact trading process and the

derivation of the price vector. Instead, we assume that the market is *perfectly competitive* and, therefore,

$$(1-\lambda_i)m_i(t) = p(t)y_i(t).$$

Roughly the argument is as follows: if the left-hand side $\geq$ the right-hand side, then it is not optimal to produce because cost is higher than revenue. Again, if the right-hand side $\geq$ the left-hand side, then there exists a *supernormal* profit which attracts more agents to produce more. But that leads to a fall in price and hence the economy comes to the equilibrium only when the left-hand side = the right-hand side. Summing up the above equation over all agents, one gets $\sum_i(1-\lambda_i)m_i(t) = p(t)\sum_i y_i(t)$. One can rewrite this as

$$M(t)V(t) = p(t)Y(t), \tag{6.19}$$

where $M(t)$ is the total money in the system and $V(t)$ is equivalent to the velocity of money at time t. It is evident that $V(t)$ depends on the parameter of the utility functions λ_i for all agents. It may be noted that the derived equation is analogous to the Fisher equation of 'quantity theory of money'. For an alternative interpretation of the Fisher equation in the context of CC-type exchange models, see Wang and Ding (2005).

(3) In this closed economy, no money is either created or destroyed during the exchange process. After all trading is done, each agent has whatever he saved for future consumption and the interest income earned from it added to some fraction $\alpha_i(t)$ of the total amount of money invested in production of the current consumption, i.e.

$$m_i(t+1) = \lambda_i m_i(t) + \alpha_i(t)\sum_i(1-\lambda_i)m_i(t),$$

or, from Eq. (6.19),

$$m_i(t+1) = \lambda_i m_i(t) + \alpha_i(t)p(t)Y(t). \tag{6.20}$$

Assume that $\alpha_i(t)p(t)Y(t) = \epsilon(t)$ to get the following reduced equation:

$$m_i(t+1) = \lambda_i m_i(t) + \epsilon(t). \tag{6.21}$$

6.6 Steady-state distribution of money and price

Each agent's money follows the dynamics defined by Eq. (6.21), in which $\epsilon(t)$ can be assumed to be a *white noise*. It can be easily shown that this process produces a gamma function–like part (we have discussed it in detail in Chapter 5) with a power law tail (Basu and Mohanty 2008). Equation (6.20) is its more general version.

This is an autoregressive process of order 1 with $\lambda_i < 1$, assuming that the last term is a white noise. Taking expectation for the whole expression

$$[1 - \lambda_i]\langle m_i \rangle = \langle \alpha_i(t) \rangle \langle p(t) \rangle \langle Y(t) \rangle.$$

Denoting $\langle \alpha_i(t) \rangle \langle p(t) \rangle \langle Y(t) \rangle$ by a finite constant C, we rewrite the equation (without the agent index) in terms of average money holding

$$\lambda = \left(1 - \frac{C}{m}\right),$$

which immediately shows that $d\lambda \propto dm/m^2$. Since $P(m)dm = \rho(\lambda)d\lambda$, where $P(m)$ is the distribution of money and $\rho(\lambda)$ is the distribution of λ, it follows that

$$P(m) = \rho\left[\left(1 - \frac{C}{m}\right)\right]\frac{1}{m^2}. \tag{6.22}$$

Such autoregressive-type equations have been proven to be very useful for mimicking the data-generating processes of the kinetic exchange models. In particular, Basu and Mohanty (2008) provide examples of the emergence of gamma function–like behaviour in money distribution for a variety of noise terms with the assumption of $\lambda_i = \lambda_j$ for all i and j. Also, as has been shown above, if λ is distributed uniformly the distribution of money has a power law feature with the exponent 2 (see also Chatterjee and Chakrabarti 2007b; Basu and Mohanty 2008). Writing Eq. (6.19) without the time index, we get

$$p = \frac{V}{Y/M}.$$

One observes that M is of the order of N, the number of agents, and Y is their total production. If we assume that both V and Y/M are distributed uniformly, one can show that the distribution of price is a power law:

$$f(p) \sim p^{-2}. \tag{6.23}$$

So, in the above model, price may also have power law fluctuations. However, there is no clear evidence supporting the existence of a power law in commodity price fluctuation, although it has been verified in stock price fluctuations (Sornette 2004).

6.7 Discussion

It is still an open problem to interpret the kinetic exchange model within the framework of the standard utility maximization used in neoclassical economics. What we have presented above is a simple general equilibrium framework. However, it is perfectly possible that it can be interpreted better by some other models. In our model, it is assumed that all agents are characterized by Cobb–Douglas-type preferences with stochastic parameters (Eq. 6.1), which they maximize subject to their respective budget constraints (Eq. 6.3). Assuming binary trades between randomly chosen agents, we have shown that the money evolution equations are just the same as in the kinetic exchange models. An important part of the result is that the money evolution equations are independent of other endogenous variables such as prices and also of the exogenous variables such as the production level. The equivalence of the money evolution equations derived in this framework with those developed earlier employing the entropy maximization principle is certainly noteworthy.

But there are a number of problematic issues associated with the theory developed so far. We treated money as a commodity which has no storage cost. The important roles of money in our model are that it is an asset which transfers purchasing power to the future. However, money has some other features as well which we ignored. For example, money in a real economy acts as a medium of exchange and also serves as a unit of accounting. But in our model we did not consider those issues. Another major point is that we did not consider debt in the model. Also, for the sake of clarity, it may be mentioned that the distributions derived so far are concerned with money only. Strictly speaking, there is no income (no wage earned from labour supply or no rental income) in these models and neither is there any wealth accumulation (no capital stock). The simple reason is that the production side is completely ignored in this model. This may be considered as a future direction of research to consider a model of production and to derive income/wealth distribution directly from that framework. From a technical point of view, the usage of only one type of utility function, namely Cobb–Douglas, is also questionable. However, even though there are many other types of utility functions in the literature, we used it because it directly translates the money evolution equation into those obtained in the kinetic exchange models.

However, the essential nature of both income and wealth and their distributions are captured very well by this class of models. In short, in this class of models money (asset) works as a proxy for income/wealth. Since the distributions derived for money compare extremely well with the empirical data of income/wealth, we believe that these models provide important insights for income/wealth distributions. Critics Gallegati *et al.* (2006) noted that 'in industrialized capitalist economies, income is most definitely not conserved', which is certainly true. But while this observation is correct that income and wealth in an economy grows over time, it does not contradict the models presented above. The growth of income and wealth over time is, by definition, a time-series phenomenon, whereas the models presented here try to explain cross-section data, i.e. the data taken at a single instance or within a very short period of time. The main argument in favour of the kinetic exchange models is that millions of small transactions that take place in a very short span of time can generate the essential stochastic features of the kinetic exchange models and the corresponding distributions. The crucial difference between the above-mentioned model and the usual modelling exercises in macroeconomics is that we did not consider the foresight of the buyers and hence there is no rational expectation about the future. However, an interesting point captured in the model is that it leaves room for contingency in the decision-making process, which is almost always ignored in the standard macroeconomic models. Also, the non-trivial interactions between the agents in the kinetic exchange model are not present in the usual representative agent-based macroeconomic models. It would be interesting to see whether these features can be meaningfully incorporated in the mainstream macroeconomic models.

7

Dynamics: generation of income, inequality and development

In some of the earlier chapters, we have presented in detail the models which have been inspired from physical theories. In the last chapter, we presented a simple economic model which intends to capture the basic features of the kinetic exchange models. The objective of this chapter is to present analytical discussions on stochastic and related economic models of the distribution of income and wealth.

7.1 The economic significance

In order to get an idea about the distributional effects of a particular economic policy it often becomes necessary to have information on the distribution of income. Also inequality based on distribution of income has important effects on development, social outcomes and public finance. The shape of the income distribution in a country enables the policy-makers to get an idea about the amount of tax collection. For instance, in Germany 50.6% (19.7%) of the entire income tax is paid by the top 10% (1%) of the taxpayers (see Merz *et al.* 2005).

There are economic and non-economic reasons for separate study of the distribution of wealth that cannot be interpreted as human capital, such as educational background. Examples of wealth of this type are financial assets and real properties. Unlike human capital they can be traded in appropriate markets at the time of necessity; for instance, when the flow of regular income reduces (after retirement) and when consumption is likely to increase with an increase in family size. Precautionary motive for saving is also regarded as a major reason for accumulation of wealth. Wealth accumulation is often taken as an indicator of social status. Wealth may be retained as well for the purpose of bequests.

A variety of size distributions, including income, wealth, employment and firms ranked in terms of assets have been observed to demonstrate many similar characteristics. These common features have led to speculation about the processes that generate such distributions. A number of possibilities have been explored in this context. One of the most popular approaches relies on the theory of probability and the models considered along this line become stochastic. There is randomness in the changes in size, but the underlying process that gives rise to the random jumps fulfils the condition that it is possible to derive the probability distribution of future size. Prediction of the size distribution of the variable under consideration (say, income or wealth) becomes possible when the same process applies to a sufficiently large number of individuals.

Of particular interest in this context are ergodic processes which are characterized by a steady-state or stationary distribution, and over time every distribution converges to this distribution. The steady-state distribution is referred to as the equilibrium distribution. If the observed frequencies can be closely approximated by steady-state frequencies, then the underlying stochastic formulation can be assumed to provide an explanation of the observed frequency distribution. In other words, the observed system is in a small neighbourhood of the equilibrium so that the observed distribution can be regarded as equilibrium distribution. In fact, it has been found in various studies that the distributions of income and wealth are characterized by some globally stable and robust features (Yakovenko and Barkley Rosser 2009). There is now a huge literature in which an economy can be viewed as a thermodynamic system and the distribution of income among economic agents is identified with the distribution of energy among particles in a gas (see Chatterjee *et al.* 2005b; Chakrabarti *et al.* 2006).

7.2 Analysis of income distributions

Two well-known features of an empirical distribution of income are that it is positively skewed and its top tail can be approximated by a Pareto distribution, which, when the logarithm of the number of persons possessing income level more than x is plotted against the logarithm of x, produces a straight line, as we have seen in most of the figures in Chapter 2.

In his pioneering contribution, Gibrat (1931) first advanced the idea that the income growth might be explained by a stochastic process. He suggested the 'law of proportionate effect', which generates a positively skewed distribution. According to the law of proportionate effect, the change in income at any time point is a random proportion of the income at the previous time point. Thus, if x_t denotes the income of a representative individual at time t, then

$$x_t - x_{t-1} = \varepsilon_t x_{t-1}, \tag{7.1}$$

where ε_t is a stochastic component. By repeated use of (7.1) we get

$$\ln(x_t) = \ln(x_0) + \sum_{j=1}^{t} \ln\left(1 + \varepsilon_j\right), \tag{7.2}$$

where x_0 is the initial income. Now, if $\ln(x_0)$ and the sequence $\{\ln(1+\varepsilon_j)\}$ are assumed to be independent and identically distributed with mean μ and variance σ^2, then by the central limit theorem for large t, x_t follows a log-normal distribution with parameters $(t+1)\mu$ and $(t+1)\sigma^2$, that is, $\ln(x_t)$ follows a normal distribution with mean $(t+1)\mu$ and variance $(t+1)\sigma^2$ (see Aitchison and Brown 1957).

The density function of $\ln(x_t)$ is then given by

$$f\left(x_t, (t+1)\mu, (t+1)\sigma^2\right) = \frac{1}{x_t\sqrt{2\pi\sigma^2(t+1)}}\exp\left[-\frac{[\ln x_t - (t+1)\mu]^2}{2\sigma^2(t+1)}\right]. \tag{7.3}$$

For the log-normal distribution $\sigma\sqrt{t+1}$ is a sensible measure of income inequality (see Chakravarty 1990). Thus, as a model of income distribution the formulation given by (7.1) has a major technical disadvantage: it predicts that income inequality, as measured by the standard deviation of the logarithm of income, increases over time unboundedly. This implication of the law of proportionate effect is not supported by evidence.

Kalecki (1945) argued that income inequality is determined by economic factors to a large extent, and there is no tendency for an increase in inequality of the type described above. He studied the case where the variance of $\ln(x_j)$ is kept constant over time by economic factors. From Eq. (7.1) it then follows that there is a negative correlation between $\ln(x_{j-1})$ and $\ln(1+\varepsilon_j)$. He assumed further that the correlation is described by an equation of the type

$$\ln(1+\varepsilon_j) = -\alpha_j \ln(x_{j-1}) + \varepsilon_j^*, \tag{7.4}$$

where α_j is a constant and ε_j^* is independent of $\ln(x_{j-1})$. This equation (7.4) may be regarded as an instrument that influences stabilization of the variance. The resulting generating equation then becomes

$$x_j = x_{j-1}^{1-\alpha_j}\exp(\varepsilon_j^*). \tag{7.5}$$

Under certain conditions the distribution of x_t can be approximated by a log-normal distribution. In the same paper, Kalecki (1945) also investigated the cases where the variance changes over time and changes can be subdivided into systematic and random components.

A variant of the law of proportionate effect was considered by Champernowne (1953). The Champernowne approach is an application of the Markov model, which states that there exist probability distributions for individual incomes in the current period, given incomes in the earlier period. Incomes are measured in intervals and there is a set of transition probabilities, each of which indicates that an income in interval i in the current period will be in interval j in the next period. This assumption is a replacement of the equal growth assumption in the law of proportionate effect. Given these transition probabilities, under certain fairly general conditions the income distribution will converge to an equilibrium distribution, which is independent of the original distribution. In different models of this type, the equilibrium distributions have been found to be some variants of the Pareto distribution (e.g. Wold and Whittle 1957).

To discuss the Champernowne model more explicitly, let us suppose that income as a size measure above a certain minimum level x_0 is divided into an infinite number of non-overlapping intervals/classes. The i-th class given by (x_{i-1}, x_i) satisfies the assumption that $x_i = c\,x_{i-1}$, where $c > 1$ is a constant. Since $\ln(x_i) - \ln(x_{i-1}) = \ln(c)$, it follows that on a logarithmic scale the end points of the size classes are equidistant. We regard the development of income over time through these intervals as a stochastic process.

Let $p_{i,j}(t)$ be the probability that an income in class i at time t will be in class $(i + j)$ at time $(t + 1)$. Then

$$\sum_{j=-(i-1)}^{\infty} p_{i,j}(t) = 1. \tag{7.6}$$

This ensures that an income in size class i at time t will be in one of the classes $1, 2, \ldots, \infty$ with probability 1. The probability that at time t an income will be in size class i is denoted by $p_i(t)$. The size distribution of income $p_i(t + 1)$ is generated by

$$p_i(t+1) = \sum_{l=-\infty}^{i-1} p_{i-l}(t)\, p_{i-l,l}(t)\,. \tag{7.7}$$

Since Eq. (7.7) relates the size distribution of income at time $(t + 1)$ with the distribution at time t through transition probabilities $p_{i,j}(t)$, it is referred to as the transition equation.

The Champernowne model is based on the assumptions C1–C3 stated below.

Assumption C1: the number of incomes is constant over time.

Assumption C2: for each i and t, and for a fixed positive integer n,

$$p_{ij}(t) = 0 \quad \text{if} \quad j > 1 \quad \text{or} \quad j < -n, \tag{7.8}$$

and

$$p_{ij}(t) = a_j > 0 \quad \text{if} \quad j > i \quad \text{and} \quad -n \leq j \leq 1. \tag{7.9}$$

Assumption C3:

$$\sum_{j=-n}^{1} j a_j < 0. \tag{7.10}$$

Assumption C1 is self-explanatory. In order to ensure validity of this assumption, it is supposed that, in the event of an income recipient's death between two time periods, there will be an heir to her income in the next period. According to Eq. (7.8) of assumption C2, no income can move up by more than one interval or move down by more than n intervals during a particular period. Equation (7.9) of assumption C2 demands that the transition probabilities are independent of time and the initial size interval i. They depend only on the number of intervals j an income can move in a single period. Assumption C3 is a stability condition. It says that there will be a tendency for incomes to shrink in the long run, that is, that incomes cannot increase unlimitedly. There will be no equilibrium distribution without such a condition.

Let us denote the equilibrium distribution by $p_i\ (e)$. Then by (7.7) and (7.9) we have

$$p_i\ (e) = \sum_{l=-n}^{1} p_{i-l}\,(e)\,a_l \tag{7.11}$$

for $i > 1$ and $p_1(e) = \sum_{l=-n}^{0} p_{1-l}\,(e)\,h_l$, where $h_l = \sum_{v=-n}^{l} a_v$. Determination of $p_i(e)$ will require determination of a solution of (7.11). For this purpose, if we substitute $p_i(e) = \bar{c}b^i$ into (7.11), we get the following polynomial equation of degree $(n + 1)$, where $\bar{c} > 0$ is a constant:

$$\Psi\,(b) = \sum_{j=-n}^{1} b^{1-j} a_j - b = 0. \tag{7.12}$$

Thus, $p_i\ (e) = \bar{c}b^i$ will solve (7.11), whenever b is a positive real root, other than unity, of Eq. (7.12).

Descartes' rule of sign changes indicates that $\Psi\,(b) = 0$ has two positive real roots. One root is unity. For determining the other root, we note that $\Psi'\,(1) = -\sum_{j=-n}^{1} ja_j$, where Ψ' stands for the derivative of Ψ. In view of the stability condition, $\Psi'\,(1) > 0$. Since $\Psi\,(0) = a_1 > 0$, $\Psi\,(1) = 0$ and $\Psi'\,(1) > 0$, the second root d, say, will satisfy the inequality $0 < d < 1$. Then

$$p_i(e) = \hat{c}d^i, \tag{7.13}$$

where $\hat{c} > 0$ is a constant such that the sum of the probabilities becomes 1.

It is now necessary to establish a connection between the probabilities in (7.13) and the underlying distribution of income. Let $F : [x_0, \infty) \to [0, 1]$ be the cumulative distribution function of income ($F\,(y)$ is the proportion of persons with income less than or equal to y). If F is assumed to be continuously differentiable, then the continuous function f, the derivative of F, is called the income density function.

Thus,

$$F(y) = \int_{x_0}^{y} f(v)\, dv. \tag{7.14}$$

Then

$$\begin{aligned} 1 - F(x_k) &= \sum_{i=k}^{\infty} p_i(e) \\ &= \hat{c} \sum_{i=k}^{\infty} d^i \\ &= \frac{\hat{c} d^k}{(1-d)}. \end{aligned} \tag{7.15}$$

This gives

$$\ln(1 - F(x_k)) = \ln\left(\frac{\hat{c}}{1-d}\right) + k \ln(d) . \tag{7.16}$$

Since $x_i = c\, x_{i-1}$ for all i, it follows that $x_k = c^k x_0$, where x_0 is the minimum income. This gives $-k \ln(c) = \ln(\frac{x_0}{x_k})$. Let $\alpha = -\frac{\ln(d)}{\ln(c)}$. Since $0 < d < 1$ and $c > 1$, $\alpha > 0$. Then we can rewrite (7.16) as

$$\ln(1 - F(x_k)) = \ln\left(\frac{\hat{c}}{1-d}\right) + \alpha \ln\left(\frac{x_0}{x_k}\right). \tag{7.17}$$

Equation (7.17) is the logarithmic form of the Pareto law under the assumption that $\ln\left(\frac{\hat{c}}{1-d}\right) = 0$. If N_x is the number of incomes exceeding the income level x, then Pareto's law says that $N_x = (1 - F(x)) N_{x_0} = Ax^{-\alpha}$, where $A > 0$ is a constant. The slope of the cumulative frequency distribution, when drawn on a double log scale, is $-\alpha$. This is in fact the strong form of the Pareto law. This form has been sometimes found to be unjustified empirically. The weak form of the law says that the curve $(\ln(N_x), \ln(x))$ should be asymptotic to the straight line which represents the strong form of the law (Mandelbrot 1960). It is often referred to as the 'power law of income'. However, the strong form has established itself as a popular income distribution model through its long usage (e.g., Rutherford 1955; Simon 1955; Reed 2001).

The Pareto density is then given by

$$f(x_k) = \alpha x_0^{\alpha} x_k^{-\alpha-1}. \tag{7.18}$$

The mean of this distribution is $\mu(F) = \frac{\alpha x_0}{(\alpha-1)}$. Thus, for the distribution to have a finite mean we need the restriction that $\alpha > 1$. The parameter α is nothing but the Pareto exponent, which can be interpreted as an inequity parameter. To see this, note that the Gini index – the most popular index of income inequality – for the Pareto distribution is

$$G(F) = 1 - \frac{x_0 + \int_{x_0}^{\infty} (1 - F(y))^2 \, dy}{\mu(F)} = \frac{1}{2\alpha - 1}. \tag{7.19}$$

As mentioned in Chapter 2, the Gini index has a natural geometric interpretation as twice the area enclosed between the diagonal line of equality and the Lorenz curve of the income distribution, where the Lorenz curve of an income distribution represents the share of the total income enjoyed by the bottom z $(0 \leq z \leq 1)$ proportion of the population and the curve coincides with the line of equality when the income distribution is perfectly equal. Thus, the greater the divergence of the curve from the line of equality, the higher the Gini index.

An income distribution-based welfare function is defined as an increasing function of the mean income and a decreasing function of inequality (see Shorrocks 1988; Amiel and Cowell 2003; Chakravarty 2009). The most popular form of the Gini welfare function is

$$Q_G(F) = \mu(F)(1 - G(F)) = \frac{2\alpha x_0}{(2\alpha - 1)} \tag{7.20}$$

(see Blackorby and Donaldson 1978; Foster and Sen 1997). Since an increase in the value of α decreases the Gini welfare function for the Pareto distribution, interpretation of α as an inequity parameter is quite reasonable.

The distinct pattern of the income distribution in the middle and in the upper tail has been noted by many researchers, including Singh and Maddala (1976). They have derived a functional form for income density based on the concept of hazard or failure rate, which is used extensively in reliability theory. For an income distribution function with density function f and distribution function F, the failure rate is defined as

$$Z(x) = \frac{f(x)}{(1 - F(x))}. \tag{7.21}$$

When time is taken as the random variable, we do not expect in most of the situations a particular benefit to accrue with time to reduce the failure rate. When the random variable is changed from time to income, a decreasing failure rate after a certain point is obvious. While more income may help in earning more, ageing is unlikely to confer any advantage unambiguously for survival or decreasing failure rate.

Often it might be convenient to consider the failure rate of a transformation of the random variable. Consider the transformation $\ln(x)$. The failure rate with respect to this transformation is then defined as $Z^*(\ln(x)) = \frac{dF/d\ln(x)}{1-F}$. The measure Z^* is called the proportionate failure rate. It can be shown that for the Pareto distribution Z^* is a constant and it is increasing for the log-normal distribution.

Singh and Maddala (1976) assumed that Z^* first increases at an increasing rate, then with a decreasing rate and then asymptotically reaches constancy. Building on this idea they developed the following form of the distribution function:

$$F(x) = 1 - \frac{1}{(1 + c_1 x^{c_2})^{c_3}}, \tag{7.22}$$

where c_1, c_2 and c_3 are positive constants. F is increasing, for $x = 0$, $F = 0$ and as $x \to \infty$, $F \to 1$. For $c_3 = 1$, F becomes the distribution suggested by Fisk (1961).

Montroll and Shlesinger (1982) indicated that the upper tail of an income distribution is well described by the Pareto (power) law and the rest follows a log-normal distribution. As already presented in detail in Chapter 2, Drăgulescu and Yakovenko (2001a) parameterized the income distribution by exponential distribution for the middle part and by the Pareto distribution for the upper tail (see also Nirei and Souma 2007). In fact, numerous studies during the last 10 years or so demonstrated that the upper tail is clearly described by the Pareto law, but the remaining part follows either a log-normal or a gamma distribution (Chakrabarti and Marjit 1995; Levy and Solomon 1997; Ispolatov *et al.* 1998; Moss de Oliveira *et al.* 1999; Aoyama *et al.* 2003; Di Matteo *et al.* 2004; Clementi and Gallegati 2005a,b; Coelho *et al.* 2005; Sinha 2006; Chakrabarti and Chakrabarti 2010b).

The density function of a random variable following the gamma distribution with parameters $\tilde{c}$ and η is given by

$$f(x) = \frac{1}{\Gamma(\eta)} (\tilde{c}x)^{\eta-1}\, \tilde{c} \exp(-\tilde{c}x), \quad x \geq 0, \tag{7.23}$$

where $\tilde{c} > 0$ is a scale parameter, $\eta > 0$ is the skewness parameter and Γ represents the gamma function defined as

$$\Gamma(\eta) = \int_0^\infty \exp(-v)\, v^{\eta-1} dv. \tag{7.24}$$

Salem and Mount (1974) fitted the gamma distribution to personal income data in the USA for the years 1960–9. It was found that the gamma distribution fits better than the log-normal.

As discussed in Chapter 6, Chakrabarti and Chakrabarti (2010b) considered a simple microeconomic model with a large number of agents and analysed the corresponding asset transfer equations arising from trading among the agents. It is shown that this type of asset transfer has close similarity with the process of energy transfer resulting from collisions among particles in a thermodynamic system like an ideal gas. As found by Gibbs, the steady-state distribution for such a system turns out to be exponential (see Yakovenko and Barkley Rosser 2009). Under certain modifications the model produces a gamma-type distribution as the distribution of money among the agents. A characterization of the power law for the upper tail of the distribution is also developed.

McDonald (1984) considered a general framework which facilitates a comparison of alternative models. The generalized gamma (GG) and generalized beta of the first and second kind (GB1, GB2) considered by McDonald are defined as

$$\left.\begin{aligned} g_1(y;\varphi,\rho,\vartheta) &= \frac{\varphi\, y^{\varphi\vartheta-1}\exp\left(-\frac{y}{\rho}\right)^{\varphi}}{\rho^{\varphi\vartheta}\Gamma(\vartheta)}, \quad y\geq 0,\\ g_2(y;\varphi,\zeta,\vartheta,q) &= \frac{\varphi\, y^{\varphi\vartheta-1}(1-(y/\zeta)^{\varphi})^{q-1}}{\zeta^{\varphi\vartheta}B(\vartheta,q)}, \quad 0\leq y\leq\zeta,\\ g_3(y;\varphi,\zeta,\vartheta,q) &= \frac{\varphi\, y^{\varphi\vartheta-1}}{\zeta^{a\vartheta}B(\vartheta,q)(1+(y/\zeta)^{\varphi})^{\vartheta+q}}, \quad 0\leq y, \end{aligned}\right\} \tag{7.25}$$

where $B(\vartheta,q)=\int_0^1 v^{\vartheta-1}(1-v)^{q-1}\,dv$ is the beta function.

For $\varphi=1$, $g_2(y;\varphi,\zeta,\vartheta,q)$, that is, GB1 becomes the beta distribution of the first kind (B1) with three parameters, considered by Thurow (1970). GB1 coincides with GG in the limiting case $q\to\infty$, under the assumption that $\zeta=\rho(\vartheta+q)^{1/\varphi}$. The beta distribution of the second kind (B2) drops out as a special case of GB2, that is, $g_3(y;\varphi,\zeta,\vartheta,q)$, when $\varphi=1$. B1 and B2 approach the gamma distribution in the limiting case $q\to\infty$ under the respective assumptions that $\zeta=\rho(\vartheta+q)$ and $\zeta=q\rho$. The GG distribution also coincides with the gamma distribution if $\varphi=1$. If we assume that $\rho^{\varphi}=\sigma^2\varphi^2$ and $\vartheta=\frac{\varphi\mu+1}{\rho^{\varphi}}$, then GG coincides with the log-normal distribution. For $\vartheta=1$, GB2 becomes the Singh–Maddala distribution, which for $\zeta=\rho q^{1/\varphi}$ produces the Weibull distribution as $q\to\infty$. Both Weibull and gamma distributions become the exponential distribution for $\varphi=\vartheta=1$.

McDonald (1984) used US family nominal income for 1970–80 for the purpose of estimation and comparison of alternative distributions. The generalized beta of the second kind was found to provide the best fit. The Singh–Maddala distribution, which facilitates estimation and analysis of results, provided a better fit than the generalized beta distribution of the first kind and all of the two- and three-parameter models.

Esteban (1986) introduced the concept of income share elasticity as a useful tool for describing the size distribution of income. Let $M(x,x+v)$ be the share of total income enjoyed by individuals with incomes in the interval $[x,x+v]$. The income share elasticity at x, $\Pi(x)$, of a given distribution is the limit when $v\to 0$

of the proportional change of $M(x, x+v)$ with respect to x, that is,

$$\Pi(x) = \lim_{v \to 0} \frac{d \ln (M(x, x+v))}{d \ln (x)}. \tag{7.26}$$

In other words, income share elasticity represents the proportionate change of income share of individuals from a certain income position as a fraction of the proportionate change of income when the change in income is infinitesimally small. Esteban showed that there exists a one-to-one correspondence between income share elasticity and density function. He developed a characterization of the GG distribution using the hypotheses that income share elasticities have a constant rate of decline and a weaker form of the weak Pareto law. This weaker version of the weak Pareto law demands that the total income at each income level falls at a constant rate. As Esteban noted for many well-known income density functions, including Pareto, log-normal, gamma and Weibull, income share elasticities have constant declining rates (see also Chakravarty and Majumder 1990).

The stochastic nature of a Champernowne-type model is subject to much criticism because of its lack of economic content. In such a model economic theory has been replaced by the element of chance. As Lydall (1968) stated, '... too much reliance is placed on the laws of chance and too little on specific factors which are known to influence the distribution'. However, some authors, including Chipman (1974), believe that economic factors have no significant role in the analysis of the distribution of income.

Chakravarty and Ghosh (2010) considered an economic approach to derive the size distribution of income. Given that the aggregate demand x in the economy consists of consumer demand $C(x)$ and investment demand $S(x)$ (saving), they looked for the distribution of income that maximizes aggregate saving when the economy meets the following restrictions: (1) the mean income and (2) social welfare are given a priori. The second restriction provides information on distributional equity. Thus, the resulting distribution is the distribution of income of a given total, on a specific indifference surface, that maximizes total saving. Here (forced) saving can be regarded as an instrument available in the hands of a social planner.

The welfare function considered by Chakravarty and Ghosh (2010) is the Donaldson and Weymark (1980, 1983) S-Gini social welfare function $Q_\delta(F)$, given by

$$Q_\delta(F) = \int_0^\infty \delta x\,(1 - F(x))^{\delta-1}\, f(x)\, dx, \tag{7.27}$$

where F is the cumulative distribution function, $f(x) = F'(x)$ is the income density function and where $\delta > 1$ is a flexibility parameter. The welfare function Q_δ coincides with the Gini welfare function when $\delta = 2$. It is assumed that the marginal propensity to consume $C'(x) \in (0, 1)$ and the average propensity to consume $C(x)/x$ is decreasing in x. Since income is either consumed or saved, $S'(x) \in (0, 1)$ and $S(x)/x$ is increasing in x. The aggregate saving can then be written as

$$S(F) = \int_0^\infty S(x) f(x) dx. \tag{7.28}$$

Denoting the given level of welfare by ϖ and the mean income by μ, the restrictions imposed by the economy can be written as

$$\int_0^\infty x f(x) \, dx = \mu \tag{7.29}$$

and

$$\int_0^\infty \delta x \, (1 - F(x))^{\delta-1} f(x) \, dx = \varpi. \tag{7.30}$$

Then, given these restrictions, the income distribution that maximizes aggregate saving has the following density function:

$$f(x) = \hat{a}\delta^{-\frac{1}{\delta-1}} S''(x) \left[\hat{a}\left(\hat{b} - S'(x)\right)\right]^{\frac{2-\delta}{\delta-1}}, \tag{7.31}$$

where $S''(x) < 0$ and $\hat{a} < 0$ and $\hat{b}$ are constants.

If we assume that the saving function is of the form

$$S(x) = A + Bx + (1 - B) x_0^\alpha \frac{x^{1-\alpha}}{1-\alpha}, \quad 0 < x_0 \leq x, \tag{7.32}$$

where $A < 0$ and B, $\frac{1}{2} < B < 1$ are constants. For this saving function, the income density function in Eq. (7.31) for $\delta = 2$ becomes

$$f(x) = \alpha x_0^\alpha x^{-\alpha-1}, \quad x \geq x_0 > 0, \tag{7.33}$$

which is the Pareto density function (same form as in Eq. (2.2)). We thus have an economic theoretic characterization of the Pareto distribution. Given A, B and x_0, an increase in α reduces the level of aggregate saving. The reason behind this is that, given concavity of the saving function, for every increment in income a smaller amount will be saved the higher is the value of α. An increase in the value of α makes the saving function more concave. There is thus a trade-off between the size of maximal aggregate saving and concavity of the saving function.

Lydall (1959) argued that there are different types of income that are distributed in different manners, particularly; difference between the distributions may arise from different sources. The Champernowne model is quite plausible in the context of income from ownership of capital. He developed a model that explains the distribution of employment income.

There are a finite number of income grades arranged in increasing order. Let n_i be the number of employees in grade i. It is assumed that $\frac{n_i}{n_{i+1}} = k$, where $k > 0$ is a constant. This assumption says that there is a fixed ratio between the number of employees in each grade and the number of employees in the grade immediately above. The income of an employee grade i is denoted by x_i. It is further assumed that $\left[\frac{x_{i+1}}{kx_i}\right] = q^*$, where $q^* > 0$ is a constant. We can rewrite this assumption as $\left[\frac{n_{i+1}x_{i+1}}{n_i x_i}\right] = q^*$. That is, the ratio between total incomes in two consecutive grades is a constant.

Let w_i denote the slope of a line connecting two adjacent points $(\ln(n_i), \ln(x_i))$ and $(\ln(n_{i+1}), \ln(x_{i+1}))$. Then

$$w_i = \frac{\ln(n_{i+1}) - \ln(n_i)}{\ln(x_{i+1}) - \ln(x_i)} = -\frac{\ln(k)}{\ln(kq^*)}. \tag{7.34}$$

Since both k and q^* are constants, w_i is a constant. Thus, the curve connecting the successive points will be a straight line. From the estimate of Allen (1957) for the distribution of higher income salaries using the UK income salary data for the year 1954–5, the plot of cumulative income frequencies on a double-logarithmic scale was close to a straight line with a slope of approximately -2.5.

Given that $-\frac{\ln(k)}{\ln(kq^*)}$ is a constant, we have $\lambda = \frac{\ln(k)}{\ln(kq^*)}$, a constant, from which it follows that $\ln(kq^*) = \ln(k)^{\frac{1}{\lambda}}$. This in turn implies that

$$kq^* = (k)^{\frac{1}{\lambda}}. \tag{7.35}$$

Thus, kq^* varies inversely with λ. This implies that proportionate salary differentials between two adjacent groups will decrease as λ increases and vice versa. As Lydall (1959, p. 114) pointed out, this finding is consistent with the interpretation that λ can be regarded 'as a measure of the degree of inequality of distribution'.

In order to specify the relation between an individual's income, her age and transition proportions, Hartog (1976) considered a Markov-chain approach to income distribution. The central idea underlying the Hartog model is that individuals are endowed with specific capabilities of which a price per unit is determined through supply and demand conditions in the labour market. Individual income is derived from capability endowment and price per unit. His formulation also indicates how these variables relate to an individual's age.

To discuss the Hartog model analytically, assume that the relation between an individual's age a and her income $y \geq 0$ is given by $y = J(a)$, where J is increasing in its argument. That is, income is assumed to be an increasing function of age. We can write the inverse of this relation as $a = J^{-1}(y) = a(y)$, say. Here $a(y)$ represents the age at which income level y is reached. We denote the stable age density function by $h(a)$. It is assumed that the age–income relationship holds for all individuals. Then the income density function Φ can be obtained from the age density function through the following transformation:

$$\Phi(y) = h(a(y)) \left| \frac{da}{dy} \right|. \tag{7.36}$$

Assume that there are K income intervals $[v_{i-1}, v_i)$, $i = 1, 2, \ldots, K$. Thus, the length of interval i is $v_i - v_{i-1}$. It is assumed further that movement of income takes place once a year. An individual with income $y \in [v_{i-1}, v_i)$ will leave interval i if the increase in y during the year is sufficient to ensure that the increased income is above v_i. We define a variable z_i as the increase in income during the year such that at the end of the year income becomes v_i, where $v_i - z_i \geq v_{i-1}$. All incomes $y \in [v_{i-1}, v_i)$ satisfying the inequality $y > v_i - z_i$ will be in the interval $[v_i, v_{i+1})$ next year.

The proportion of individuals with incomes in interval $[v_{i-1}, v_i)$ who can reach interval $[v_i, v_{i+1})$ next year is

$$r_i = \frac{\int_{v_i - z_i}^{v_i} \Phi(y)\, dy}{\int_{v_{i-1}}^{v_i} \Phi(y)\, dy}. \tag{7.37}$$

Given that there is a monotonic relationship between income and age, we can also express Eq. (7.37) using the age density function as follows:

$$r_i = \frac{\int_{a(v_i - z_i)}^{a(v_i)} h(a)\, da}{\int_{a(v_{i-1})}^{a(v_i)} h(a)\, da}. \tag{7.38}$$

Thus, while Eq. (7.37) represents the proportion of individuals whose incomes move from an interval to the next higher interval at the end of the year, Eq. (7.38) gives the proportion of individuals whose ages are such that their incomes grow into the next higher interval in 1 year.

By our assumption the age difference required for an annual income difference z_i is 1 year. This implies that $a(v_i - z_i) = a(v_i) - 1$. Consequently, we can rewrite Eq. (7.38) as

$$r_i = \frac{\int_{a(v_i)-1}^{a(v_i)} h(a)\, da}{\int_{a(v_{i-1})}^{a(v_i)} h(a)\, da}. \tag{7.39}$$

The formulation we have considered so far did not take into account mortality – more generally, the withdrawal of individuals from the labour force. Let $m > 0$ be the constant mortality rate per unit of time. That is, m is assumed to be independent of age. This then corresponds to a probability $p(a)$ of surviving up to age a,

$$p(a) = \exp(-ma), \tag{7.40}$$

and the rate of departure within a year of $1 - \exp(-m)$. This probability and rate of departure are independent of income intervals. Upward mobility of individuals in interval $[v_{i-1}, v_i)$ then equals $[1 - (1 - \exp(-m))]r_i = \exp(-m)r_i$.

Since the age density has been assumed to be stable, we can restrict it to the form $h(a) = \upsilon \exp(-g^* a)\, p(a), \upsilon \geq 0$ being the birth rate and g^* the rate of growth of the labour force. Letting $\upsilon = (m + g^*)$ we can write the age density function as

$$h(a) = \upsilon \exp(-\upsilon a). \tag{7.41}$$

In order to illustrate the age–income profile, we denote an individual's stock of labour at age a by l. Let $\bar{l}$ be the maximum number of units of labour the individual can supply. Then,

$$l = \bar{l}\left[1 - \exp(-\gamma a)\right], \quad \gamma > 0. \tag{7.42}$$

Thus, the process of accumulation of experience is an increasing function of age. In the limit it converges to $\bar{l}$, the rate of convergence of being γ. Support for this type of specification of accumulation of experience can be found in the literature on learning curves (Corlett and Morecombe 1970). The two parameters $\bar{l}$ and γ may vary across occupations.

The individual labour supply function is defined as $L = sl$, where $0 \leq s \leq 1$ is the proportion of stock of labour that the individual decides to supply. This proportion depends on the wage rate per unit of capability, ω. A similar analysis holds for the demand side of the labour market. We denote the equilibrium wage rate by ω^*. The age–income profile then becomes

$$y = \bar{y}[1 - \exp(-\gamma a)], \tag{7.43}$$

where $\bar{y} = \omega^* s(\omega^*)\, \bar{l}$. Thus, given the equilibrium wage rate, $\bar{y} = \omega^* s(\omega^*)\, \bar{l}$ gives the total amount of income that the individual can earn when he has the maximum level of experience. The rate at which income converges to its maximum level equals that of experience.

To determine the income distribution, note from (7.43) that we can write the inverse age–income profile as

$$a = -\frac{1}{\gamma} \ln\left[1 - \frac{y}{\bar{y}}\right]. \tag{7.44}$$

Differentiation of Eq. (7.44) and substitution of the result and Eq. (7.44) into the age density function in Eq. (7.36) yields the following form of the income density function

$$\Phi(y) = \frac{\upsilon}{\gamma}(\bar{y})^{\frac{-\upsilon}{\gamma}}[\bar{y} - y]^{\frac{\upsilon}{\gamma}-1}. \tag{7.45}$$

The shape of the income distribution is fully determined by the parameters υ and γ, where υ is the birth rate of the population and at the same time represents the decay of the age density. On the other hand, γ is the rate of convergence of individual income towards its maximum attainable value $\bar{y}$. The slope of the income density is negative or positive, corresponding to $\upsilon > \gamma$ or $\upsilon < \gamma$. If $\upsilon = \gamma$, the income density function is a constant.

We solve Eq. (7.39) using Eqs. (7.41) and (7.44), and obtain the transition proportion as follows:

$$r_i = (\exp(\upsilon) - 1)\left[\left(\frac{\bar{y} - \upsilon_{i-1}}{\bar{y} - \upsilon_i}\right)^{\frac{\upsilon}{\gamma}} - 1\right]^{-1}. \tag{7.46}$$

The Hartog model clearly establishes a relationship between age–income profiles, income distribution and the Markov-chain approach. Given the age–income profile, the distribution of personal income is derived under the assumptions that there are individuals at different points on the same age–income profile. The model also discusses mobility of individuals through the income distribution.

7.3 Analysis of wealth distributions

The Champernowne model analysed in the earlier section can also be regarded as a model of wealth distribution. A number of authors, including Steindl (1972), Shorrocks (1975) and Vaughan (1979), have also suggested alternative types of stochastic models that give rise to distributions of wealth with a Pareto-type asymptotic upper tail. Fiaschi and Marsili (2010) considered a general equilibrium model with a large number of firms and dynasties interacting through the capital and labour markets and characterized the equilibrium distribution of wealth. It has been shown that the upper tail of the distribution can be well represented by a Pareto distribution, where the underlying parameter depends on the rate of return on capital, rate of savings, population growth rate, tax on capital and portfolio diversification. On the other hand, the lower part of the distribution depends on the labour market functioning (see also Atkinson and Harrison 1978).

As an illustrative example of a stochastic model, we briefly analyse the Shorrocks model, which relies on queuing theory. Consider an individual with a non-negative stock of wealth at any time point. Assume that within the time interval $(t, t + \Delta t]$ of length Δt an individual with j units of wealth has the probability $(d_1 j + d_2)\,\Delta t + \mathcal{O}(\Delta t)$ of receiving one more unit and a very small probability of possessing more than one unit, where $\mathcal{O}(\Delta t)$ represents the terms that tend to zero faster than Δt. Likewise, the probability of reduction of wealth by one unit over the same time interval of length Δt is given by $d_3 j\,\Delta t + \mathcal{O}(\Delta t)$ and the probability of losing more than one unit is negligible. From the assumptions about these probabilities, it is clear that the changes in current wealth holding at a rate proportional to the holding level itself are influenced by parameters d_1 and d_3, which may be identified with capital gains and losses, unearned income, etc. On the other hand, the parameter d_2 might be influenced by earned income.

The three basic assumptions of the model are that: (1) the law of proportionate effect holds, (2) the process is a Markov process and (3) the parameters of the process are constant. According to assumption (2), the probability of a change in the size depends on the current size and is independent of past history. Assumption (3) is a time homogeneity assumption.

Let $p_j(t)$ denote the probability of an individual possessing j units of wealth at time t. The set of Kolmogorov forward equations that describe the process is given by

$$\left.\begin{aligned} \frac{dp_j(t)}{dt} &= -\left[(d_3+d_1)\,j+d_2\right]p_j(t)+\left[d_1(j-1)+d_2\right]p_{j-1}(t) \\ &\quad + d_3(j+1)\,p_{j+1}(t), \quad j \geq 1, \\ \frac{dp_0(t)}{dt} &= -d_2 p_0(t) + d_3 p_1(t). \end{aligned}\right\} \tag{7.47}$$

If an individual starts with i units at the initial period (time 0), then the initial conditions are $p_i(0) = 1$, $p_j(0) = 0$, $j \neq i$. We write $\pi(v,t) = \sum_{j=0}^{\infty} p_j(t)\,v^j$ for the probability-generating function. Then

$$\left.\begin{aligned} \frac{\partial \pi(v,t)}{\partial t} &= \sum_{j=0}^{\infty} \frac{dp_j(t)}{dt} v^j, \\ \frac{\partial \pi(v,t)}{\partial v} &= \sum_{j=0}^{\infty} j p_j(t)\,v^{j-1}. \end{aligned}\right\} \tag{7.48}$$

Substituting the equations from Eq. (7.47) into Eq. (7.48) we get

$$\frac{\partial \pi(v,t)}{\partial t} - (d_3 - d_1 v)(1-v)\frac{\partial \pi(v,t)}{\partial v} = -d_2(1-v)\,\pi(v,t). \tag{7.49}$$

The general solution of this partial differential equation is given by

$$\pi(v,t) = (d_3 - d_1 v)^{-\frac{d_2}{d_1}}\,\psi\left(\frac{(1-v)\,e^{(d_1-d_3)t}}{d_3 - d_1 v}\right), \tag{7.50}$$

where the arbitrary function ψ is determined from the initial conditions. For instance, if an individual starts with i units at the initial period, then $\pi(v, 0) = v^i$ and

$$\psi(y) = \left(\frac{d_3 y - 1}{d_1 y - 1}\right)^i \left(\frac{d_1 - d_3}{d_1 y - 1}\right)^{\frac{d_2}{d_1}}. \tag{7.51}$$

It is also true that

$$1 = \sum_{j=0}^{\infty} p_j(t) = \pi(1, t) = (d_3 - d_1)^{-\frac{d_2}{d_1}} \psi(0). \tag{7.52}$$

Consequently, if $d_3 > d_1$

$$Lt_{t\to\infty}\pi(v, t) = (d_3 - d_1 v)^{-\frac{d_2}{d_1}} \psi(0) = \left(\frac{d_3 - d_1}{d_3 - d_1 v}\right)^{\frac{d_2}{d_1}} = \pi^*(v). \tag{7.53}$$

This is the probability-generating function at the equilibrium or stationary-state distribution. Therefore, in this ergodic process, irrespective of initial wealth holding, the probability distribution of individual wealth holding will converge over time to $\pi^*(v)$.

If the same process is assumed to hold for a large number of individuals, then $p_j(t)$ can be interpreted as the proportion of individuals owning j units of wealth at time t. The original wealth distribution is then represented by the initial probability-generating function $\pi(v, 0)$, and, under the assumption $d_3 > d_1$, π^* emerges as the equilibrium distribution in the sense that over time the distribution converges to π^*. Expanding $\pi^*(v)$ as a power series of v and determining the coefficients of the terms, equilibrium frequencies of the proportion of individuals owning j units of wealth are given by

$$p_j^* = \frac{\left(1 - \frac{d_1}{d_3}\right)^{\frac{d_2}{d_1}} \Gamma\left(\frac{d_2}{d_1} + j\right)}{\Gamma\left(\frac{d_2}{d_1}\right)\Gamma(j+1)} \left(\frac{d_1}{d_3}\right)^j \quad j = 0, 1, \ldots. \tag{7.54}$$

This is a negative binomial distribution and, as $j \to \infty$,

$$p_j^* \to A_1 j^{\frac{d_2}{d_1} - 1} \left(\frac{d_1}{d_3}\right)^j, \tag{7.55}$$

where $A_1 > 0$ is a constant. Under the assumption that $\left(\frac{d_1}{d_3}\right)$ can be approximated by unity and $\left(\frac{d_2}{d_1}\right) < 1$, Simon (1955) suggested this as a form representing the upper tail of size distributions. This shows some similarity of the stationary state to empirically observed distributions.

However, the model has certain limitations. For instance, in order to harmonize it with the dynamic features of empirical distributions one possibility is to allow the states of the system to change over time. But an equilibrium system defined on variable units may give rise to some growth features and a suitable adjustment may be unsatisfactory. If a timeless state is assumed, then we have to allow the parameters of process to vary. As Shorrocks (1975) demonstrated, if the parameters are subject to variation and the equilibrium distribution changes, there may not be eventual convergence of the variance and the coefficient of variation at time t to their corresponding stationary-state values. The mean at time t converges to its equilibrium value monotonically at a very low speed.

It is clear that the argument of Lydall (1968) regarding the lack of economic content in stochastic models applies here as well. If the process parameters are taken to be functions of economic variables, then the time homogeneity assumption becomes incompatible. Shorrocks (1975) also considered a variant of this model by dropping the time homogeneity assumption. Under certain mild conditions, the probability-generating function of the equilibrium corresponding to the parameter values at time t becomes

$$\pi^* (v, t) = \left(\frac{d_3 (t) - d_1 (t) v}{d_3 (t) - d_1 (t)} \right)^{\frac{-d_2(t)}{d_1(t)}}. \tag{7.56}$$

Note that this distribution can vary over time. The sequence of actual distributions $\pi (v, t)$ may not approach the equilibrium distribution $\pi^* (v, t)$ and there is no a-priori reason to expect convergence. The absolute deviation between means at time t and at equilibrium and also the absolute deviation between corresponding variances may increase monotonically over the lifetime of the process. By relating the parameters to economic variables, the system gives us an opportunity to incorporate economic theory into stochastic models.

We now present some economic theoretic approaches to the analysis of size distribution of wealth. Parts of our presentation are based on Davies and Shorrocks (2000). We begin by analysing the simple framework advocated by Meade (1964, 1975), which is based on the following accounting identity for W_i, wealth in period i:

$$W_i = W_{i-1} + E_i + \tau_i W_{i-1} - C_i + I_i, \tag{7.57}$$

where C_i and E_i are, respectively, consumption and earned income in period i, net of taxes and transfers, τ_i is the (average) net rate of return on investment and I_i stands for net inheritances (gifts and bequests) received in period i. If inheritances are assumed to be incorporated into the initial wealth W_0 and consumption depends on both income and wealth, then

$$\begin{aligned} W_i &= W_{i-1} + \varsigma_i (E_i + \tau_i W_{i-1}) - \xi_i W_{i-1} \\ &= (1 + \varsigma_i \tau_i - \xi_i) W_{i-1} + \varsigma_i E_i \\ &= (1 - \Omega_i) W_{i-1} + \varsigma_i E, \end{aligned} \tag{7.58}$$

where ς_i represents the average rate of saving from current income and ξ_i is the fraction of wealth accumulated in period $(i-1)$ spent on consumption in period i and $\Omega_i = \xi_i - \varsigma_i \tau_i$ denotes the 'internal rate of deaccumulation out of wealth'. If the average rate of return increases with the level of wealth, then it is likely that the differences in individual rates of return will have a disequalizing influence (Meade 1964). Clearly, this equation can be interpreted in an intergenerational framework, where W_i stands for the lifetime wealth of generation i and Ω can be assumed to capture the impact of wealth taxation and bequest splitting (Atkinson and Harrison 1978). In a sense, the Meade framework also absorbs the insights of the life cycle behaviour, which we analyse below.

Modigliani and Brumberg (1954) pioneered the life cycle saving model of intergenerational wealth accumulation, which is closely related to the permanent income hypothesis of Friedman (1957). There are now several variations of the model. But all of them retain the following basic assumptions: (1) the consumers are assumed to be forward looking and their preferences are defined over present and future consumptions, and possibly leisure period; and (2) it is expected that life will end with a period of retirement. The simplest form of the model assumes that everybody faces the same rate of interest χ, which is taken to be constant. It is assumed further that there is no uncertainty and no bequest motive, and everyone has the same length of life T. Denote a representative consumer's consumption and income earned in period i by C_i and E_i, respectively. The consumer then maximizes the intertemporal utility function

$$U(C_1, C_2, \cdots C_T) \tag{7.59}$$

subject to the constraints

$$C^L = \sum_{i=1}^{T} \frac{C_i}{(1+\chi)^{i-1}} \leq \sum_{i=1}^{R} \frac{E_i}{(1+\chi)^{i-1}} = E^L, \tag{7.60}$$

where R is the date of retirement, and C^L and E^L denote, respectively, lifetime consumption and lifetime earnings. If it is necessary to ensure that saving should be undertaken for the purpose of financing consumption in retirement, then appropriate restrictions can be incorporated into the functional form of U.

Now assume that the intertemporal utility function is additively separable,

$$U = \sum_{i=1}^{T} \frac{u(C_i)}{(1+\phi)^{i-1}}, \tag{7.61}$$

where ϕ is the rate of time preference. Although a more general formulation is to allow variability of the function u over time, constant time preference ensures consistent consumption planning over time (Strotz 1956). Various functional forms can be assumed for u. For instance, we may make a specific assumption about the risk aversion measure of the individual with utility function u, where a risk aversion measure enables us to judge to what extent a person is risk averse. If u displays a constant absolute risk aversion $-\frac{u''}{u'}$, then $u(C_i) = e^{-\tilde{\kappa} C_i}$, where $\tilde{\kappa} > 0$ is the constant absolute risk aversion level. On the other hand, if u has a constant relative risk aversion $-\frac{C_i u''(C_i)}{u'(C_i)}$, then

$$u(C_i) = \begin{cases} \dfrac{C_i^{1-\kappa}}{1-\kappa}, & \kappa > 0, \kappa \neq 1, \\ \ln(C_i), & \kappa = 1, \end{cases} \tag{7.62}$$

where $\kappa > 0$ is the coefficient of constant relative aversion. If we incorporate the forms given by Eq. (7.61) and Eq. (7.62) into Eq. (7.60), then optimal consumption of the individual satisfies

$$C_{i+1} = \left(\frac{1+\chi}{1+\phi}\right)^{\frac{1}{\kappa}} C_i = (1+g)\, C_i, \tag{7.63}$$

where $g \cong \frac{\chi-\phi}{\kappa}$ is the constant rate of growth of planned consumption. The intertemporal elasticity of substitution here is given by $\frac{1}{\kappa}$. See also Wang (2007) for an analysis of wealth distribution using constant relative risk averse utility-based precautionary savings demand. For an earlier treatment, see Carroll (1997).

Several implications of the prediction of constant growth rate of consumption can be examined. If it is assumed that earnings are constant up to retirement and zero after that, and interest rate and planned consumption growth rate are zero, then the consumer will save a fixed amount annually before retirement and dissave a constant amount annually after retirement, with accumulated saving being zero at the time of death. This gives rise to a hump-shaped age profile of wealth which increases linearly with age until retirement and then decreases linearly to zero. Clearly, the pick of the profile occurs in a close neighbourhood of the age of retirement. There can be high wealth inequality between the richest (in the neighbourhood of the retirement age) and the poorest (those who have just begun working lives and those who are not far from death).

More general implications of the consumption path (7.63) can be studied if we assume that $g \cong \frac{\chi-\phi}{\kappa} > 0$, that is, planned consumption grows at a positive constant rate. For a typical hump-shaped age profile individuals dissave at young age, save fairly in the middle age and dissave in old age (in retirement). However, it has also been argued that at young age most individuals are net savers and if family size remains constant on retirement consumption decreases at an increasing rate, producing less dissaving than that suggested by the model (Attanasio 1998). If we incorporate labour–leisure choice in the model and goods and leisure are assumed to be substitutes, then it is highly likely that on retirement goods will be substituted by leisure, which in turn leads to a downward jump in consumption (Davies 1988).

In order to look at more implications of the model considered above we assume that $\chi = \phi = 0$, where consumption in period i equals wealth plus expected future earnings as a proportion of the length of the remaining lifetime,

$$C_i = \frac{W_{i-1} + \mathbb{E}\left(\sum_{j=1}^{R} E_j\right)}{T - i + 1}, \tag{7.64}$$

where $\mathbb{E}$ stands for the expectation operator. The two components of earnings are permanent and transitory: $E_t = E_t^p + \varepsilon_t^*$, where the transitory component ε_t^* is a disturbance term with zero mean and finite variance. It has been observed that the earnings process in the USA can be approximated by a combination of white noise and a random walk (Hubbard *et al.* 1994; Carroll 1997). This gives $\mathbb{E}\left(E_j^p\right) = E_i^p, j = i, \ldots, R$. Consequently,

$$C_i = \frac{W_{i-1} + (R - i + 1) E_i^p + \varepsilon_i^*}{T - i + 1}. \tag{7.65}$$

It follows that the propensity to consume out of permanent earnings is higher than that out of current wealth or transitory component. There is a tendency for earnings inequality to grow over the working lifetime. The wealth built up by saving out of past permanent incomes will be more unequal. If transitory earnings are not simply measurement errors, then high propensity to save out of transitory income along with the condition that everybody starts life with zero wealth will imply that the transitory income as a source of wealth inequality decreases over time (Davies and Shorrocks 2000).

It is possible to incorporate uncertain lifetime into the life cycle model. Under the constant relative risk aversion assumption the objective function becomes

$$\mathbb{E}(U_i) = \sum_{j=i}^{T} \frac{\left(1 - q_j\right) C_j^{1-\kappa}}{(1-\kappa)(1+\phi)^{j-i}}, \tag{7.66}$$

where T now stands for the maximum length of life and q_j denotes the probability of death before period j. If the nature of insurance markets is such that there exist fair annuities, then there will be complete annuitization of wealth by all individuals, where a fair annuity is a promise to pay a fixed amount of money each year for a certain period. In this case the objective function in Eq. (7.66) will be maximized subject to the constraint

$$\sum_{j=i}^{T} \frac{C_j\left(1-q_j\right)}{(1+\chi)^{j-i}} \leq W_{i-1} + \sum_{j=i}^{R} \frac{E_j\left(1-q_j\right)}{(1+\chi)^{j-i}} . \tag{7.67}$$

Yaari (1965) demonstrated that in this case also planned consumption follows the constant growth path given by Eq. (7.63).

In practice the annuity market may be imperfect (Friedman and Warshawsky 1988). In this case the objective function in (7.66) is to be maximized subject to the constraint (7.60) and a non-negative wealth constraint. This solution then produces a non-constant growth rate of consumption given by $\frac{\chi-\phi-m_i}{\kappa}$, where $m_i = \frac{q_{i+1}-q_i}{1-q_i}$ is the probability of death of an individual in period $(i+1)$ given that he has survived up to period i. The predicted age profile of wealth in this case turns out to be different from that which emerged in the case certainty. If preferences are assumed to display constant relative risk aversion, then for sufficiently small $\chi - \phi$ the constant growth pattern can give rise to a realistic hump-shaped age profile of consumption or a profile with a predominant hump shape and an increasing trend in later years of retirement.

Models of wealth distributions dealing with intergenerational issues generally incorporate demographic factors such as marriage and fertility, and economic factors. In several models of intergenerational transfers, there has been explicit specification of parental preferences. In the Becker and Nigel (1976) altruistic model the parental preferences are represented by

$$V = V\left(c^p; c_1, c_2, \ldots, c_N\right), \tag{7.68}$$

where c^p is parental lifetime consumption, c_i is the lifetime consumption of child i and N is the number of children. Parents provide a bequest b_i to child i and their human capital investment on this child i is l_i. Child i's earning e_i is the product of her human capital H_i, which is a function of l_i, and human capital rental rate w:

$$e_i = wH_i(l_i), \quad i = 1, 2, \ldots, N. \tag{7.69}$$

Let y^p be the parental lifetime income. Then assuming that the interest rate is zero, the consumptions of parents and the children are given by

$$\left.\begin{aligned} c^p &= y^p - \sum_{i=1}^{N} (l_i + b_i), \\ c_i &= e_i + b_i, \quad i = 1, 2, \ldots, N. \end{aligned}\right\} \tag{7.70}$$

The parent's objective is then to maximize the utility function

$$V = V\left(y^p - \sum_{i=1}^{N} (l_i + b_i); e_1 + b_1, \ldots, e_N + b_N\right), \tag{7.71}$$

with respect to b_i's and l_i's subject to the constraints $b_i \geq 0$ and $l_i \geq 0$, $i = 1, 2, \ldots, N$.

If capital market is assumed to be perfect, then it is possible for parents to borrow against their children's earnings. The latter is a relaxation of the condition $b_i \geq 0$. The optimization problem can then be decomposed into two parts. The first part requires the choice of an efficient level of investment in each child, l_i^*. This will generate a corresponding level of earning for child i, e_i^*. In the next part of the problem, Eq. (7.71) is maximized with respect to b_i's, assuming that e_i^*'s are given. Under the assumption that all parents treat the children symmetrically, it follows that

$$e_1^* + b_1 = \cdots = e_N^* + b_N. \tag{7.72}$$

This is perfect equalization of children's incomes net of transfers. Thus, in this altruistic model bequests play an important role in reducing equilibrium inequality. It is possible to avoid this conclusion if we do not drop the assumption $b_i \geq 0$ (Atkinson 1988; Wilhelm 1996).

Bequests may be zero if parental income is low. If parental income and children's ability to repay for investments in their human capital were perfectly correlated, then it is likely that bequests would be zero up to the threshold income of parents and then there would be an increasing relationship. This possibly provides an explanation of the finding that income elasticity of bequests is less than 1 for a large bottom part of the population and greater than 1 for the remaining part (Menchik and Martin 1983). The model also explains how the behaviour of parents may lead to a reduction in the degree of intergenerational income mobility.

Benhabib and Bisin (2006) studied the dynamics of the distribution of wealth in an overlapping generations economy with bequests and different forms of redistributive taxation. The economy is assumed to be populated by a continuum of individuals with a constant probability of death. A dead individual is substituted by her child, which means that the population is stationary. A fraction of the individuals has a 'joy of giving' bequest motive towards its children. Individuals are born with an initial wealth which consists of the bequests of their parents and welfare subsidy from the government, if any. The interest rate is constant. At any time an individual allocates her wealth between an asset and an annuity. The main qualitative characteristics of the stationary wealth distribution are skewedness and fat tails. In particular, it is shown that the stationary wealth distribution is a Pareto law. Levy (2003) formulated a general stochastic process of wealth accumulation in terms of capital investment and investigated the conditions under which convergence to the empirically observed Pareto distribution is ensured. De Nardi (2004) used the 'joy of giving' bequest motive and death rate to increase the wealth concentration. Benhabib and Zhu (2008) made use of two kinds of uncertainty – investment risk and death risk – to produce the wealth distribution with Pareto tails.

The 'joy of giving' bequest motive along with portfolio selection was also assumed in a continuous time overlapping generations economy considered by Zhu (2010). It is shown that idiosyncratic investment risk gives rise to a Pareto tail in the wealth distribution. Simulation results demonstrate that the wealth distribution has a fat tail and produces a Gini index and Lorenz curve close to the US wealth distribution, where the period of investigation was 2001–7. Quadrini (2000) and Cagetti and De Nardi (2009) introduced entrepreneurships into the heterogeneous agent model and matched the fat tail of wealth distribution.

Benhabib *et al.* (2011) investigated the impact of the inheritance of the investment ability on wealth inequality. It is shown that the wealth distribution has a fat tail and the tail index was used to characterize the fatness of the tail. They considered an economy composed of households that live for T periods. At each time point t, households of any age in the interval $[0, T]$ are alive. A household born at time Θ has a child entering the economy at time $(T + \Theta)$ after her parents' death. Household generations are overlapping with the link from dynasties. A household taking birth at time Θ belongs to the $i = \left(\frac{\Theta}{T}\right)$-th generation of its dynasty. In this economy post-tax 'joy of giving' bequests of parents are initial wealth of children. Thus, if W_{i+1} denotes the initial wealth of an i-th generation household, then it is shown that W_{i+1} follows the process

$$W_{i+1} = X_{i+1} W_i + Y_{i+1}, \tag{7.73}$$

where X_{i+1} and Y_{i+1} are stochastic processes that represent, respectively, the rate of return on wealth across generations and the permanent income of a generation. If X_{i+1} and Y_{i+1} are independent and identically distributed, then this dynamics of wealth converges to a stationary distribution satisfying the strong Pareto law.

Even in the case where X_{i+1} and Y_{i+1} are correlated, it is shown that the stationary wealth distribution has a Pareto tail.

Thus in this chapter, we have considered many economic models and interpretations of the inequality in income and wealth distributions, and the different forms of the Pareto law.

8
Outlook

> Any city, however small, is in fact divided into two, one the city of the poor, the other of the rich; these are at war with one another.
>
> *Plato, ancient Greek philosopher (427–347 BC)*

Throughout the recorded history of human civilization, we have witnessed the bitter outcomes of economic inequality – social tensions, conflicts, etc. This incessant problem has been addressed by some of the greatest thinkers, philosophers and social scientists, including economists. Questions on the nature of the distributions of wealth and income have been raised repeatedly. More so, during or just after periods of crisis, wars and social calamities. In this book, we have tried to present a new interdisciplinary approach in analysing and dealing with the age-old problem of economic inequality in the societies. This paradigmatic shift has been possible owing to the combined efforts of economists, mathematicians and physicists (Cockshott *et al.* 2009; Sinha *et al.* 2010).

Noting that this inequality has a very robust and universal statistical form (discussed extensively in the first two chapters of this book), and the fact that some core human ability factors, such as the intelligence quotient or health factors, are distributed according to the normal (or Poisson, at times) distribution, a natural question to ask is why are the distributions in wealth and income so different from the normal? Why do they have such broad distributions, and with ubiquitous power law tails? The very fact that these distributions have such different characteristics (and are universally observed) indicates that there must be a deeper cause (and a common underlying mechanism).

Physicists have come up with a purely physical reason for the origin of these questions: they identified the role of entropy maximization in the stochastic dynamics of the economic markets. In essence, these econophysics attempts (described in detail in this book) indicate that the origin of inequality may be more intrinsic to the market dynamics than previously envisaged.

8.1 Chapters in a nutshell

The income or wealth distribution among the population has a generic feature, as shown in Fig. 1.1. Chapter 2 gave extensive data from various time periods and across the countries, confirming the same. In this book, we have taken the position that apart from the top 5–10% of the very rich population in any country, whose numbers decay with a power law in the Pareto tail, the probability density of income for the rest of the population follows a gamma function form (and not in a numerically equivalent log-normal way, as traditionally favoured by the economists; see Eq. (1.1)).

In the next two chapters (Chapters 3 and 4), we discussed various agent-based models, which proved to be quite successful in capturing the essential empirical features of markets. As could be seen there, these efforts captured the various observed features of income or wealth distribution in societies discussed in Chapter 2. These models captured well both the initial gamma-like feature in the steady-state distributions for the income distributions of poor and middle-income group agents (below income or wealth x_c in Eq. (1.1)) and the Pareto tail of the distribution for the super-rich (with income or wealth beyond x_c in Eq. (1.2)). A noteworthy transition behaviour, discussed in Section 4.6.2, could be observed in the kinetic exchange markets, when one of the two agents involved in any exchange had to be below a preassigned threshold or 'poverty-line'. Apart from interesting social consequences, this model poses intriguing questions regarding the existence of statistical physics of phase transitions in such simple (non-interacting and ideal gas-like) kinetic exchange models.

In Chapters 4 and 5, we gave detailed numerical and analytical results for the CC (Chakraborti and Chakrabarti 2000) and CCM (Chatterjee *et al.* 2004) kinetic exchange models, which successfully captured the observed features of the income or wealth distribution in societies. As we showed in the earlier chapters, the socialist norm of *equal* income distribution ($P(x) = \delta(x - x_0); x_0 = X/N$) was quite unstable with respect to any conceivable dynamics of income or wealth exchange among the agents. In particular, as shown in Chapters 4 and 5, the entropy maximizing kinetic exchange dynamics lead to:

(1) the exponentially decaying Gibbs distribution, when the agents did not save in any trade or exchange (DY model);
(2) the gamma-like distribution Eq. (1.1), when each agent saved a (uniform) fixed fraction of its instantaneous money or wealth (CC model);
(3) Pareto power law distribution Eq. (1.2), when each agent saved but the saving fractions differed from agent to agent (CCM model).

In Chapter 6, we showed that a Cobb–Douglas-type utility maximizing dynamics readily gives the same dynamics as in the CC model, when agents have the same saving propensity (the generalization to non-uniform saving propensity case, as in CCM model, is straightforward). The equivalence of the money evolution equations derived in this framework with those developed employing the entropy maximization principle is certainly noteworthy.

The general observation, following the CCM model, is that, in spite of dispersions in income or wealth, the average income or wealth of the agents in the Pareto tail increases with their saving propensity. Some might propose that there is an apparent contradiction: the rich people in society are also those who usually take maximum risk, whereas increased value of saving propensity usually means increasing risk aversion tendency! However, in the context of the CCM model, this contradiction is resolved: in a population with mixed saving propensities, the trade-to-trade income fluctuation increases with the saving propensity value (see Fig. 4.6); the sustainability of this increased income fluctuation by the richer agents can be interpreted as the result of the attitude towards increased risk.

The dynamics of inequality were discussed in Chapter 6, as well as in Chapter 7, and a more formal economic treatise was given. However, it must be mentioned that this is not by any means a complete description and some general criticisms remain to be addressed.

8.2 Beyond income and wealth

The analyses of the income and wealth distributions and the kinetic exchange models of markets developed in this connection, and discussed in this book, have already been extended to various other social distribution and interaction phenomena.

8.2.1 Probability distribution of energy consumption

A rapid technological development of human society since the industrial revolution has been based on consumption of fossil fuel, such as coal, oil and natural gas, accumulated inside the Earth for billions of years. The physical standards of living in modern society are primarily determined by the level of per capita energy consumption. Now we understand that these fuel reserves will be exhausted in the not too distant future. In addition, the consumption of fossil fuel releases carbon dioxide into the atmosphere, which is a major greenhouse gas, affecting global climate – a global problem posing great technological and social challenges (Rezai *et al.* 2012).

There is a huge variation in the per capita energy consumption around the globe. This heterogeneity is a challenge and complicates the situation for arriving at a global consensus on how to deal with the energy issues. It has become necessary to understand the origin of this global inequality in energy consumption and

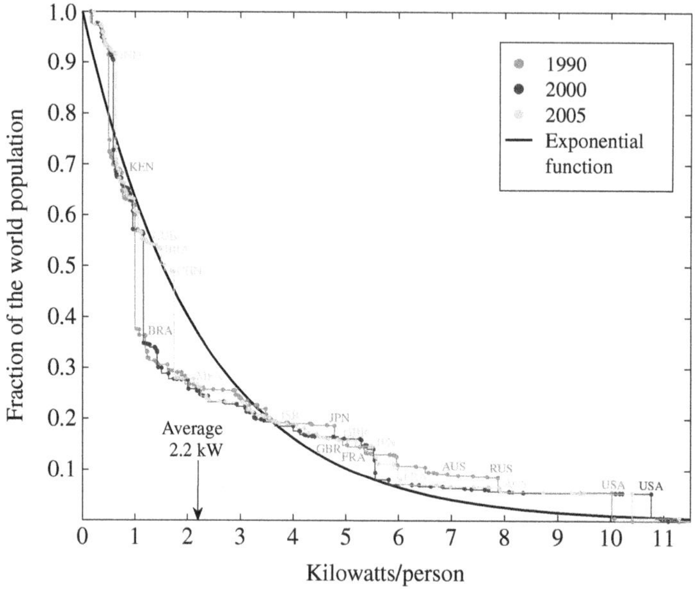

Figure 8.1 World distribution of energy consumption. Cumulative distribution functions $C(\epsilon)$ for the per capita energy consumption around the world for 1990, 2000 and 2005. The solid curve is the exponential function (8.1) with the parameter $T_\epsilon = \langle \epsilon \rangle = 2.2$ kW. Reproduced from Banerjee and Yakovenko (2010).

characterize it quantitatively (Banerjee and Yakovenko 2010; Yakovenko 2012). Banerjee and Yakovenko (2010) approach this problem using the method of entropy maximization.

Banerjee and Yakovenko (2010) and Yakovenko (2012) consider an ensemble of economic agents j, each characterized by the energy consumption ϵ_j per unit time. It is important to note that ϵ_j denotes not energy but power, measured in kilowatts (kW). They introduce the probability density $P(\epsilon)$, so that $P(\epsilon)\,d\epsilon$ gives the probability of having energy consumption in the interval $(\epsilon,\ \epsilon + d\epsilon)$. The energy production is based on the extraction of fossil fuel from the Earth, which is a physically limited resource, and is divided for consumption among the global population, and it is quite improbable to equally divide this resource. More likely, this resource is divided according to the entropy maximization principle, subject to the constraint of the global energy production. Following simple calculations (Banerjee and Yakovenko 2010; Yakovenko 2012) it was shown that $P(\epsilon)$ follows the exponential law

$$P(\epsilon) = c\, e^{-\epsilon/T_\epsilon}, \tag{8.1}$$

where c is a normalization constant, and the temperature $T_\epsilon = \langle \epsilon \rangle$ is the average energy consumption per capita. From the data of the World Resources Institute, Banerjee and Yakovenko (2010) constructed the probability distribution of per

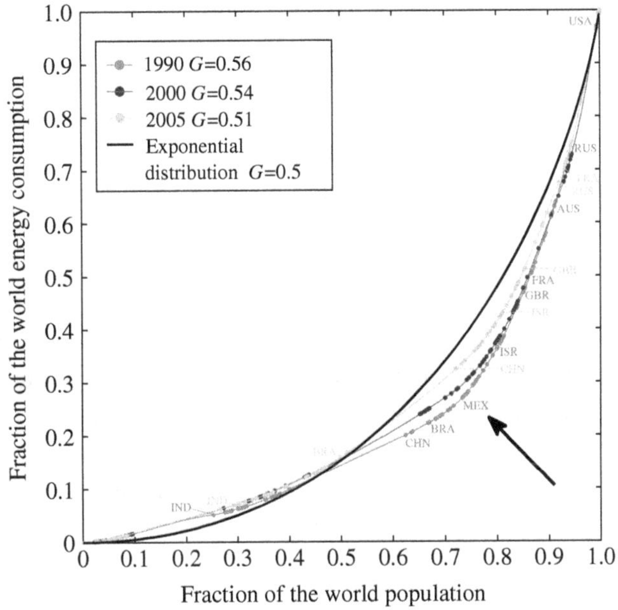

Figure 8.2 Inequality in the global energy consumption Lorenz curves for the per capita energy consumption around the world in 1990, 2000 and 2005, compared with the Lorenz curve (2.8) for the exponential distribution. Reproduced from Banerjee and Yakovenko (2010).

capita energy consumption around the world and found that it follows the exponential law approximately (Fig. 8.1). The world average energy consumption per capita is $\langle \epsilon \rangle = 2.2$ kW, compared with 10 kW in the USA and 0.6 kW in India (Banerjee and Yakovenko 2010). This means that, if India and other developing countries use the same level of per capita energy consumption as the USA, the energy resources in the world would not be sufficient to provide it. Yakovenko (2012) argues that the global energy consumption inequality is a result of the constraint on energy resources, and the global monetary inequality adjusts accordingly to implement this constraint. Owing to the fact that money is used to purchase energy, a fraction of the world population ends up being poor, and their energy consumption stays limited.

Figure 8.2 shows the Lorenz curves for the global energy consumption per capita in 1990, 2000 and 2005 from Banerjee and Yakovenko (2010). The black solid line is the theoretical Lorenz curve (2.8) for the exponential distribution (8.1). The empirical curves are reasonably close to the theoretical curve, but with some deviations. On the Lorenz curve for 1990, the arrow denotes the point where the slope of the curve changes appreciably, separating developed and developing countries. Clearly Mexico, Brazil, China and India are below, whereas the UK, France, Japan, Australia, Russia and the USA are above. The gap in the per capita energy consumption between these two groups of countries is reflected in the slope change of the Lorenz curve – the developed and developing countries differ in the degree of energy consumption and utilization, rather than in more sophisticated monetary measures. Interestingly, it is observed that the Lorenz curve for 2005 is closer to the exponential curve with a less prominent kink, indicating a reduction in the energy consumption inequality and a reduced gap between developed and developing countries, also reflected in the decrease in the Gini coefficient G listed in Fig. 8.2. Yakovenko (2012) thinks that this result can be attributed to the rapid globalization and stronger mixing of the world economy in the last 20 years. However, the distribution of energy consumption in a well-mixed globalized world economy approaches an exponential distribution, and not an equal distribution. The inherent inequality of the global energy consumption makes it difficult for the countries widely apart in the distribution to agree on consistent measures to address the energy and climate challenges. Yakovenko (2012) concludes that a transition from fossil fuel to renewable energy gives hope for achieving a global society with more equality.

8.2.2 Kinetic models of opinion formation

Some kinetic models of opinion dynamics have been proposed (Lallouache *et al.* 2010a), taking inspiration from the kinetic exchange models of wealth distributions (Chakraborti and Chakrabarti 2000; Chatterjee and Chakrabarti 2007b). The model incorporates the two-body exchange of 'opinions' between a pair of agents during a 'discussion'. The question of interest is whether, through such successive two-body discussions across the population, a consensus can be reached or not. As in many other well-known opinion models (see Castellano *et al.* 2009), they consider $o_i(t) \in [-1, +1]$ to be the opinion of an individual i at time t. In a system of N individuals, opinions evolve through binary interactions (Lallouache *et al.* 2010a; Biswas *et al.* 2011):

$$\left.\begin{aligned} o_i(t+1) &= \lambda[o_i(t) + \epsilon_t o_j(t)], \\ o_j(t+1) &= \lambda[o_j(t) + \epsilon'_t o_i(t)], \end{aligned}\right\} \tag{8.2}$$

where ϵ_t, ϵ'_t are drawn randomly from uniform distributions in [0, 1]. Here, $\lambda \in [0, 1]$ is a parameter, which is interpreted as *conviction* of an individual, or how strongly the individual retains her old opinion. Here, the *influence* factor of the counterpart in the discussion is taken to be identical to the respective conviction factor. The simple case above (Lallouache *et al.* 2010a) considers a society in which everyone has the same value of conviction and influence factors λ, which can be generalized. It is important to note that there are no conservation laws here, unlike the wealth exchange models of Chakraborti and Chakrabarti (2000) and

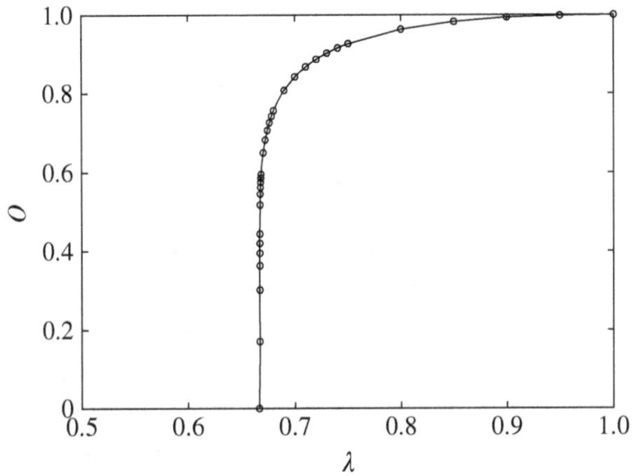

Figure 8.3 Spontaneous social consensus formation: Numerical results for the variation of the order parameter $O = |\sum_i o_i|/N$ (steady-state value) against λ, following the dynamics of Eq. (8.2).

Chatterjee and Chakrabarti (2007b). The constraint that the opinions are bounded, i.e. $-1 \leq o_i(t) \leq 1$, introduces the non-linearity in the dynamics. The system undergoes a phase transition. The ordering in the system may be measured by a quantity (order parameter) $O = |\sum_i o_i|/N$. The multiagent system (dynamics given by Eq. (8.2)) goes into either of the two possible phases: for any $\lambda \leq \lambda_c$, $o_i = 0 \; \forall i$, while for $\lambda > \lambda_c$, $O > 0$ and $O \to 1$ as $\lambda \to 1$ (Fig. 8.3). Here the critical point of the phase transition $\lambda_c \simeq 0.667$, also supported by mean-field theory ($\lambda_c = 2/3$). The details of the studies are given in Lallouache *et al.* (2010a) and Biswas *et al.* (2011). A few variants of the model have also been proposed (Sen 2010; Biswas *et al.* 2011) and a mean-field theory gives good estimates of the critical points.

8.3 Open problems and challenges

As any careful reader will realize, our attempts here have perhaps raised more open questions than we set out to answer! However, this only reinforces the statement we made in the Preface: our endeavour has been similar to fighting the monster Hydra, who grows two heads in place of an injured one! This only reflects the grandeur of the problem addressed in the book, and the versatility of our (kinetic exchange modelling) approach.

Several challenges and interesting directions have been thrust. On the science and physics side, one still seeks the closed form solutions for the distribution functions emerging out of the kinetic exchange equations (in particular for the uniform saving propensity of the agents). Also, the intriguing threshold-induced phase transition behaviour in kinetic exchanges remains to be understood. On the economics side, it would be necessary to incorporate the kinetic exchange model within the framework of the standard utility maximization used in neoclassical economics. On the policy side, the theoretical observation that increased average saving propensity of the agents in the market can decrease the dispersion in the distributions is very intriguing, and it corresponds to some indirect market observations.

We believe that the econophysics modelling of income and wealth distributions, discussed in this book, will inspire further and more intense developments in both physics and economics. This will not only help in exploring the 'natural origin' of the ubiquitous 'economic inequality', one of the most annoying and debated (philosophically as well as scientifically) issues of human history, but may also help us in identifying the potential (scientific) steps in reducing the inequality in the near future!

References

Abramowitz, M., and Stegun, I.A. 1970. *Handbook of Mathematical Functions*. New York, Dover.

Abul-Magd, A.Y. 2002. Wealth distribution in an ancient Egyptian society. *Physical Review E*, 66, 057104.

Aghion, P., Caroli, E., and Garca-Pealosa, C. 1999. Inequality and economic growth: the perspective of the new growth theories. *Journal of Economic Literature*, 37, 1615–1660.

Aitchison, J., and Brown, J.A.C. 1957. *The Lognormal Distribution*. Cambridge, Cambridge University Press.

Albert, R., and Barabási, A.-L. 2002. Statistical mechanics of complex networks. *Reviews of Modern Physics*, 74, 47–97.

Alesina, A., and Perotti, R. 1996. Income distribution, political instability, and investment. *European Economic Review*, 40, 1203–1228.

Alesina, A., and Rodrik, D. 1992. Distribution, political conflict, and economic growth: a simple theory and some empirical evidence. In Cukierman, A., Hercowitz, Z., and Leiderman, L. (eds.), *Political Economy, Growth and Business Cycles*. Cambridge, MIT Press.

Allen, R.G.D. 1957. Changes in the distribution of higher incomes. *Economica*, 24, 138–153.

Allen, M.P., and Tildesley, D.J. 1989. *Computer Simulation of Liquids*. Oxford, Oxford University Press.

Amiel, Y., and Cowell, F.A. 2003. Inequality, welfare and monotonicity. *Research on Economic Inequality*, 9, 35–46.

Angle, J. 1986. The surplus theory of social stratification and the size distribution of personal wealth. *Social Forces*, 65, 293–326.

Angle, J. 2006. The inequality process as a wealth maximizing process. *Physica A*, 367, 388–414.

Angle, J. 2010. The inequality process vs. the saved wealth model. Two particle systems of income distribution: which does better empirically? *Munich Personal RePEc Archive*. See http://econpapers.repec.org/RePEc:pra:mprapa:20835.

Aoyama, H., Nagahara, Y., Okazaki, M.P., *et al.* 2000. Pareto's law for income of individuals and debt of bankrupt companies. *Fractals*, 8, 293–300.

Aoyama, H., Souma, W., and Fujiwara, Y. 2003. Growth and fluctuations of personal and company's income. *Physica A*, 324, 352–358.

Aoyama, H., Fujiwara, Y., Ikeda, Y., *et al.* 2011. *Econophysics and Companies: Statistical Life and Death in Complex Business Networks*. New York, Cambridge University Press.

Atkinson, A.B. 1988. Comment in Chapter 5. In Kessler, D., and Masson, A. (eds.), *Modelling the Accumulation and Distribution of Wealth*. New York, Oxford University Press.

Atkinson, A.B., and Bourguignon, F. 2000. *Handbook of Income Distribution*. Amsterdam, Elsevier.

Atkinson, A.B., and Harrison, A.J. 1978. *Distribution of Personal Wealth in Britain*. Cambridge, Cambridge University Press.

Atkinson, A.B., and Piketty, T. 2007. *Top Incomes over the Twentieth Century*. Oxford, Oxford University Press.

Attanasio, O.P. 1998. *Consumption Demand*. Working Paper w6466. Cambridge, MA, National Bureau of Economic Research.

Ausloos, M., and Pękalski, A. 2007. Model of wealth and goods dynamics in a closed market. *Physica A*, 373, 560–568.

Axtell, R.L. 2001. Zipf distribution of U.S. firm sizes. *Science*, 293, 1818–1820.

Bak, P. 1996. *How Nature Works: the Science of Self-organized Criticality*. New York, Copernicus.

Banerjee, A., and Yakovenko, V.M. 2010. Universal patterns of inequality. *New Journal of Physics*, 12, 075032.

Banerjee, A., Yakovenko, V.M., and Di Matteo, T. 2006. A study of the personal income distribution in Australia. *Physica A*, 370, 54–59.

Barro, R.J. 1999. *Inequality, Growth, and Investment*. Working Paper 7038. Cambridge, MA, National Bureau of Economic Research.

Barrow, R., and Sala-i Martin, X. 2004. *Economic Growth*. Cambridge, MA, MIT Press.

Bassetti, F., and Toscani, G. 2010. Explicit equilibria in a kinetic model of gambling. *Physical Review E*, 81, 066115.

Basu, U., and Mohanty, P.K. 2008. Modeling wealth distribution in growing markets. *European Physical Journal B*, 65, 585–589.

Becker, G.S., and Nigel, T. 1976. Child endowments and the quantity and quality of children. *Journal of Political Economy*, 84, S143–S162.

ben Avraham, D., Ben-Naim, E., Lindenberg, K., and Rosas, A. 2003. Self-similarity in random collision processes. *Physical Review E*, 68, 050103(R).

Benabou, R. 1994. Human capital, inequality, and growth: a local perspective. *European Economic Review*, 38, 817–826.

Benabou, R. 2000. Unequal societies: income distribution and the social contract. *American Economic Review*, 90, 96–129.

Benhabib, J., and Bisin, A. 2006. The distribution of wealth and redistributive policies. *Working Paper*.

Benhabib, J., and Zhu, S. 2008. Age, Luck, and Inheritance. Working Paper 14128. Cambridge, MA, National Bureau of Economic Research.

Benhabib, J., Bisin, A., and Zhu, S. 2011. The distribution of wealth and fiscal policy in economies with finitely lived agents. *Econometrica*, 79, 123–157.

Bennati, E. 1988a. *La simulazione statistica nell'analisi della distribuzione del reddito: modelli realistici e metodo di Monte Carlo*. Pisa, ETS Editrice.

Bennati, E. 1988b. Un metodo di simulazione statistica nell'analisi della distribuzione del reddito. *Rivista Internazionale di Scienze Economiche e Commerciali*, 35, 735–756.

Bennati, E. 1993. Il metodo Monte Carlo nell'analisi economica. In: *Rassegna di lavori dell'ISCO X*, vol. 31.

Bhattacharya, K., Mukherjee, G., and Manna, S.S. 2005. Detailed simulation results for some wealth distribution models in econophysics. In Chatterjee, A., Yarlagadda, S., and Chakrabarti, B.K. (eds.), *Econophysics of Wealth Distributions*. New Economic Windows Series. Milan, Springer-Verlag.

Biswas, S., Chandra, A.K., Chatterjee, A., and Chakrabarti, B.K. 2011. Phase transitions and non-equilibrium relaxation in kinetic models of opinion formation. *Journal of Physics: Conference Series*, 297, 012004.

Blackorby, C., and Donaldson, D. 1978. Measures of relative equality and their meaning in terms of social welfare. *Journal of Economic Theory*, 18, 59–80.

Blau, J.R., and Blau, P.M. 1982. The cost of inequality: metropolitan structure and violent crime. *American Sociological Review*, 47, 114–129.

Bobylev, A.V. 1988. The theory of the nonlinear spatially uniform Boltzmann equation for Maxwellian molecules. *Soviet Scientific Reviews C*, 7, 111–233.

Bouchaud, J.-P., and Mézard, M. 2000. Wealth condensation in a simple model of economy. *Physica A*, 282, 536–545.

Braun, D. 2001. Assets and liabilities are the momentum of particles and antiparticles displayed in Feynman-graphs. *Physica A*, 290, 491–500.

Cagetti, M., and De Nardi, M. 2009. Estate taxation, entrepreneurship, and wealth. *American Economic Review*, 99, 85–111.

Carroll, C.D. 1997. Buffer-stock saving and the life cycle/permanent income hypothesis. *The Quarterly Journal of Economics*, 112, 1–55.

Case, K.E., and Fair, R.C. 2008. *Principles of Economics*. Pearson Education Limited.

Castellano, C., Fortunato, S., and Loreto, V. 2009. Statistical physics of social dynamics. *Reviews of Modern Physics*, 81, 591–646.

Chakrabarti, A.S., and Chakrabarti, B.K. 2009. Microeconomics of the ideal gas like market models. *Physica A*, 388, 4151–4158.

Chakrabarti, A.S., and Chakrabarti, B.K. 2010a. Inequality reversal: effects of the savings propensity and correlated returns. *Physica A*, 389, 3572–3579.

Chakrabarti, A.S., and Chakrabarti, B.K. 2010b. Statistical theories of income and wealth distribution. *Economics: The Open-Access, Open-Assessment E-Journal*, 4.

Chakrabarti, B.K., and Chatterjee, A. 2004. Ideal gas-like distributions in economics: effects of saving propensity. In Takayasu, H. (ed.), *Applications of Econophysics*. Tokyo, Springer.

Chakrabarti, B.K., and Marjit, S. 1995. Self-organisation and complexity in simple model systems: game of life and economics. *Indian Journal of Physics B*, 69, 681–698.

Chakrabarti, B.K., Chakraborti, A., and Chatterjee, A. (eds.). 2006. *Econophysics and Sociophysics: Trends and Perspectives*. Weinheim, Wiley–VCH.

Chakraborti, A. 2002. Distributions of money in model markets of economy. *International Journal of Modern Physics C*, 13, 1315–1321.

Chakraborti, A., and Chakrabarti, B.K. 2000. Statistical mechanics of money: how saving propensity affects its distribution. *European Physical Journal B*, 17, 167–170.

Chakraborti, A., and Patriarca, M. 2008. Gamma-distribution and wealth inequality. *Pramana*, 71, 233–243.

Chakraborti, A., and Patriarca, M. 2009. Variational principle for the Pareto power law. *Physical Review Letters*, 103, 228701.

Chakraborti, A., Pradhan, S., and Chakrabarti, B.K. 2001. A self-organising model of market with single commodity. *Physica A*, 297, 253–259.

Chakraborti, A., Muni Toke, I., Patriarca, M., *et al.* 2011. Econophysics review. II. Agent-based models. *Quantitative Finance*, 11, 1013–1041.

Chakraborty, A., and Manna, S.S. 2010. Weighted trade network in a model of preferential bipartite transactions. *Physical Review E*, 81, 016111.

Chakraborty, A., Mukherjee, G., and Manna, S.S. 2012. Conservative self-organized extremal model for wealth distribution. Fractals, 20, 163–177; http://arxiv.org/abs/1110.2075v1.

Chakravarty, S.R. 1990. *Ethical Social Index Numbers*. New York, Springer-Verlag.

Chakravarty, S.R. 2009. *Inequality, Polarization and Poverty: Advances in Distributional Analysis*. New York, Springer-Verlag.

Chakravarty, S.R., and Ghosh, S. 2010. A model of income distribution. In Basu, B., Chakrabarti, B.K., Chakravarty, S.R., and Gangopadhyay, K. (eds.), *Econophysics and Economics of Games, Social Choices and Quantitative Techniques*. New Economic Windows Series. Milan, Springer-Verlag.

Chakravarty, S.R., and Majumder, A. 1990. Distribution of personal income: development of a new model and its application to U.S. income data. *Journal of Applied Econometrics*, 5, 189–196.

Champernowne, D.G. 1953. A model of income distribution. *The Economic Journal*, 63, 318–351.

Champernowne, D.G., and Cowell, F.A. 1998. *Economic Inequality and Income Distribution*. Cambridge, Cambridge University Press.

Chatterjee, A. 2009. Kinetic models of wealth exchange on directed networks. *European Physical Journal B*, 67, 593–598.

Chatterjee, A. 2010. On kinetic asset exchange models and beyond: microeconomic formulation, trade network and all that. In Naldi, G., Pareschi, L., and Toscani, G. (eds.), *Mathematical Modeling of Collective Behavior in Socio-Economic and Life Sciences*. Boston, MA, Birkhäuser.

Chatterjee, A., and Chakrabarti, B.K. 2006. Kinetic exchange models with single commodity having price fluctuations. *European Physical Journal B*, 54, 399–404.

Chatterjee, A., and Chakrabarti, B.K. 2007a. Ideal-gas like market models with saving: quenched and annealed cases. *Physica A*, 382, 36–41.

Chatterjee, A., and Chakrabarti, B.K. 2007b. Kinetic exchange models for income and wealth distributions. *European Physical Journal B*, 60, 135–149.

Chatterjee, A., and Sen, P. 2010. Agent dynamics in kinetic models of wealth exchange. *Physical Review E*, 82, 056117.

Chatterjee, A., Chakrabarti, B.K., and Manna, S.S. 2003. Money in gas-like markets: Gibbs and Pareto laws. *Physica Scripta*, T106, 36–38.

Chatterjee, A., Chakrabarti, B.K., and Manna, S.S. 2004. Pareto law in a kinetic model of market with random saving propensity. *Physica A*, 335, 155–163.

Chatterjee, A., Chakrabarti, B.K., and Stinchcombe, R.B. 2005a. Analyzing money distributions in 'ideal gas' models of markets. In Takayasu, H. (ed.), *Practical Fruits of Econophysics*. Tokyo, Springer.

Chatterjee, A., Yarlagadda, S., and Chakrabarti, B.K. (eds.). 2005b. *Econophysics of Wealth Distributions*. New Economic Windows Series. Milan, Springer-Verlag.

Chatterjee, A., Chakrabarti, B.K., and Stinchcombe, R.B. 2005c. Master equation for a kinetic model of trading market and its analytic solution. *Physical Review E*, 72, 026126.

Chatterjee, A., Chakrabarti, B.K., and Sinha, S. 2007. Economic inequality: is it natural? *Current Science*, 92, 1383–1389.

Chipman, J.S. 1974. The welfare ranking of Pareto distributions. *Journal of Economic Theory*, 9, 275–282.

Clementi, F., and Gallegati, M. 2005a. Pareto's law of income distribution: evidence for Germany, the United Kingdom, the United States. In Chatterjee, A., Yarlagadda, S., and Chakrabarti, B.K. (eds.), *Econophysics of Wealth Distributions*. New Economic Windows Series. Milan, Springer-Verlag.

Clementi, F., and Gallegati, M. 2005b. Power law tails in the Italian personal income distribution. *Physica A*, 350, 427–438.

Clementi, F., Di Matteo, T., and Gallegati, M. 2006. The power-law tail exponent of income distributions. *Physica A*, 370, 49–53.

Clementi, F., Gallegati, M., and Kaniadakis, G. 2007. κ-generalized statistics in personal income distribution. *European Physical Journal B*, 57, 187–193.

Clementi, F., Di Matteo, T., Gallegati, M., and Kaniadakis, G. 2008. The κ-generalized distribution: a new descriptive model for the size distribution of incomes. *Physica A*, 387, 3201–3208.

Cockshott, P., and Cottrell, A. 2008. *Probabilistic political economy and endogenous money*. Conference on Probabilistic Political Economy, Kingston University, London. See http://www.dcs.gla.ac.uk/publications/PAPERS/8935/probpolecon.pdf.

Cockshott, W.P., Michaelson, G.J., Cottrell, A., *et al.* 2009. *Classical Econophysics*. Routledge.

Coelho, R., Néda, Z., Ramasco, J.J., and Augusta Santos, M. 2005. A family-network model for wealth distribution in societies. *Physica A*, 353, 515–528.

Coelho, R., Richmond, P., Barry, J., and Hutzler, S. 2008. Double power laws in income and wealth distributions. *Physica A*, 387, 3847–3851.

Collier, M.R. 2004. Are magnetospheric suprathermal particle distributions (functions) inconsistent with maximum entropy considerations? *Advances in Space Research*, 33, 2108–2112.

Cordier, S., Pareschi, L., and Toscani, G. 2005. On a kinetic model for a simple market economy. *Journal of Statistical Physics*, 120, 253–277.

Corlett, E.N., and Morecombe, V.J. 1970. Straightening out learning curves. *Personnel Management*, 2, 14–19.

Das, A., and Yarlagadda, S. 2003. Analytic treatment of a trading market model. *Physica Scripta*, T106, 39–40.

Davies, J.B. 1988. Family size, household production, and life cycle saving. *Annales d'Economie et de Statistique*, 9, 141–165.

Davies, J.B., and Shorrocks, A.F. 2000. *The Distribution of Wealth*, vol. 1. North-Holland Elsevier, pp. 605–675.

De Nardi, M. 2004. Wealth inequality and intergenerational links. *Review of Economic Studies*, 71, 743–768.

Derrida, B., Godréche, C., and Yekutieli, I. 1991. Scale invariant regime in the one dimensional models of growing and coalescing droplets. *Physical Review A*, 44, 6241–6251.

Desai, M.I., Mason, G.M., Dwyer, J.R., *et al.* 2003. Evidence for a suprathermal seed population of heavy ions accelerated by interplanetary shocks near 1 AU. *The Astrophysical Journal*, 588, 1149–1162.

Diaz-Giménez, J., Quadrini V., and Rios-Rull, J.V. 1997. Dimensions of inequality: facts on the U.S. distributions of earnings, income, and wealth. *Federal Reserve Bank of Minneapolis Quarterly Review*, 21, 3–21.

Di Matteo, T., Aste, T., and Hyde, S.T. 2004. Exchanges in complex networks: income and wealth distributions. In Mallamace, F., and Stanley, H.E. (eds.), *The Physics of Complex Systems (New Advances and Perspectives)*. Amsterdam, Elsevier.

Donaldson, D., and Weymark, J.A. 1980. A single-parameter generalization of the Gini indices of inequality. *Journal of Economic Theory*, 22, 67–86.

Donaldson, D., and Weymark, J.A. 1983. Ethically flexible Gini indices for income distributions in the continuum. *Journal of Economic Theory*, 29, 353–358.

Dorogovtsev, S.N., and Mendes, J.F.F. 2003a. Accelerated growth of networks. In Bornholdt, S., and Schuster, H.G. (eds.), *Handbook of Graphs and Networks*. Weinhein, Wiley-VCH.

Dorogovtsev, S.N., and Mendes, J.F.F. 2003b. *Evolution of Networks: From Biological Nets to the Internet and WWW*. Oxford, Oxford University Press.

Drăgulescu, A.A., and Yakovenko, V.M. 2000. Statistical mechanics of money. *European Physical Journal B*, 17, 723–729.

Drăgulescu, A.A., and Yakovenko, V.M. 2001a. Evidence for the exponential distribution of income in the USA. *European Physical Journal B*, 20, 585–589.

Drăgulescu, A.A., and Yakovenko, V.M. 2001b. Exponential and power-law probability distributions of wealth and income in the United Kingdom and the United States. *Physica A*, 299, 213–221.

Drăgulescu, A.A., and Yakovenko, V.M. 2003. Statistical mechanics of money, income, and wealth: a short survey. In Garrido, P.L., and Marro, J. (eds.), *Modeling of Complex Systems: Seventh Granada Lectures*, vol. 661. American Institute of Physics (AIP) Conference Proceedings. Melville, NY, AIP.

Düring, B., and Toscani, G. 2007. Hydrodynamics from kinetic models of conservative economies. *Physica A*, 384, 493–506.

Düring, B., Matthes, D., and Toscani, G. 2005. Exponential and algebraic relaxation in kinetic models for wealth distribution. In Manganaro, N., Monaco, R., and Rionero, S. (eds.), *"WASCOM 2007": Proceedings of the 14th Conference on Waves and Stability in Continuous Media*. Hackensack, NJ, World Scientific Publishing.

Düring, B., Matthes, D., and Toscani, G. 2008. Kinetic equations modelling wealth redistribution: a comparison of approaches. *Physical Review E*, 78, 056103.

Ernst, M.H., and Brito, R. 2002. High-energy tails for inelastic Maxwell models. *Europhysics Letters*, 58, 182–187.

Esteban, J.M. 1986. Income-share elasticity and the size distribution of income. *International Economic Review*, 27, 439–444.

Ferrero, J.C. 2004. The statistical distribution of money and the rate of money transference. *Physica A*, 341, 575–585.

Ferrero, J.C. 2005. The monomodal, polymodal, equilibrium and nonequilibrium distribution of money. In Chatterjee, A., Yarlagadda, S., and Chakrabarti, B.K. (eds.), *Econophysics of Wealth Distributions*. New Economic Windows Series. Milan, Springer Verlag.

Fiaschi, D., and Marsili, M. 2010. Economic interactions and the distribution of wealth. In Basu, B., Chakrabarti, B.K., Chakravarty, S.R., and Gangopadhyay, K. (eds.), *Econophysics and Economics of Games, Social Choices and Quantitative Techniques*. New Economic Windows Series. Milan, Springer-Verlag.

Fischer, R., and Braun, D. 2003a. Nontrivial bookkeeping: a mechanical perspective. *Physica A*, 324, 266–271.

Fischer, R., and Braun, D. 2003b. Transfer potentials shape and equilibrate monetary systems. *Physica A*, 321, 605–618.

Fisk, P.R. 1961. The graduation of income distributions. *Econometrica*, 29, 171–185.

Foley, D.K. 1994. A statistical equilibrium theory of markets. *Journal of Economic Theory*, 62, 321–345.

Foley, D.K. 1996. Statistical equilibrium in a simple labor market. *Macroeconomics*, 47, 125–147.

Forbes, K.J. 2000. A reassessment of the relationship between inequality and growth. *American Economic Review*, 90, 869–887.

Foster, J.E., and Sen, A. 1997. On economic inequality: after a quarter century. *On Economic Inequality*. Oxford, Clarendon Press.

Foucault, M. 2003. *Le Pouvoir psychiatrique: Cours au College de France 1973–74*. Paris, Gallimard.

Friedman, M. 1957. *A Theory of the Consumption Function*. Princeton, NJ, Princeton University Press.

Friedman, B.M., and Warshawsky, M. 1988. Annuity prices and saving behavior in the United States. In Bodie, Z., Shoven, J., and Wise, D. (eds.), *Pensions in the U.S. Economy*. Chicago, University of Chicago Press.

Fuentes, M.A., Kuperman, M., and Iglesias, J.R. 2006. Living in an irrational society: wealth distribution with correlations between risk and expected profits. *Physica A*, 371, 112–117.

Fujiwara, Y., Souma, W., Aoyama, H., Kaizoji, T., and Aoki, M. 2003. Growth and fluctuations of personal income. *Physica A*, 321, 598–604.

Gabaix, X. 1999. Zipf's law for cities: an explanation. *Quarterly Journal of Economics*, 114, 739–767.

Gallegati, M., Keen, S., Lux, T., and Ormerod, P. 2006. Worrying trends in econophysics. *Physica A*, 370, 1–6.

Garibaldi, U., and Scalas, E. 2010. *Finitary Probabilistic Methods in Econophysics*. Cambridge, Cambridge University Press.

Garlaschelli, D., and Loffredo, M.I. 2008. Effects of network topology on wealth distributions. *Journal of Physics A*, 41, 224018.

Ghosh, A., Gangopadhyay, K., and Basu, B. 2009. Consumer expenditure distribution in India, 1983-2007: evidence of a long Pareto tail. *Physica A*, 390, 83–97.

Ghosh, A., Basu, U., Chakraborti, A., and Chakrabarti, B.K. 2011. Threshold-induced phase transition in kinetic exchange models. *Physical Review E*, 83, 061130.

Gibrat, R. 1931. *Les Ingalites Economiques*. Paris, Libraire du Recueil Sirey.

Gini, C. 1921. Measurement of inequality of incomes. *The Economic Journal*, 31, 124–126.

Goswami, S., Chatterjee, A., and Sen, P. 2011. Antipersistent dynamics in kinetic models of wealth exchange. *Physical Review E*, 84, 051118.

Guala, S.D. 2009. Taxes in a simple wealth distribution model by inelastically scattering particles. *Interdisciplinary Description of Complex Systems*, 7, 1–7.

Hartog, J. 1976. Age-income profiles, income distribution and transition proportions. *Journal of Economic Theory*, 13, 448–457.

Hasegawa, A., Mima, K., and Duong-van, M. 1985. Plasma distribution function in a superthermal radiation field. *Physical Review Letters*, 54, 2608–2610.

Hayes, B. 2002. Follow the money. *American Scientist*, 90, 400.

Hegyi, G., Néda, Z., and Santos, M.A. 2007. Wealth distribution and Pareto's law in the Hungarian medieval society. *Physica A*, 380, 271–277.

Hogan, J. 2005. There's only one rule for the rich. *New Scientist*, 21 March.

Hogg, R.V., Mckean, J.W., and Craig, A.T. 2007. *Introduction to Mathematical Statistics*. Delhi, Pearson Education.

Hu, M.-B., Wang, W.-X., Jiang, R., *et al.* 2006. A unified framework for the Pareto law and Matthew effect using scale-free networks. *European Physical Journal B*, 53, 273–277.

Hu, M.-B., Jiang, R., Wu, Q.-S., and Wu, Y.-H. 2007. Simulating the wealth distribution with a richest-following strategy on scale-free network. *Physica A*, 381, 467–472.

Hubbard, R.G., Skinner, J., and Zeldes, S.P. 1994. Expanding the life-cycle model: precautionary saving and public policy. *The American Economic Review*, 84, 174–179.

Iglesias, J.R. 2010. How simple regulations can greatly reduce inequality. *Science and Culture (Kolkata, India)*, 76, 437.

Iglesias, J.R., and de Almeida, R.M.C. 2012. Entropy and equilibrium state of free market models. *European Physical Journal B*, 85, 85.

Iglesias, J.R., Gonçalves, S., Abramson, G., and Vega, J.L. 2004. Correlation between risk aversion and wealth distribution. *Physica A*, 342, 186–192.

Ispolatov, S., Krapivsky, P.L., and Redner, S. 1998. Wealth distributions in asset exchange models. *European Physical Journal B*, 2, 267–276.

Kac, M. 1959. *Probability and Related Topics in the Physical Sciences*. London, Interscience.

Kakwani, N. 1980. *Income Inequality and Poverty*. Oxford, Oxford University Press.

Kalecki, M. 1945. On the Gibrat distribution. *Econometrica*, 13, 161–170.

Kar Gupta, A. 2006a. Models of wealth distributions: a perspective. In Chakrabarti, B.K., Chakraborti, A., and Chatterjee, A. (eds.), *Econophysics and Sociophysics: Trends and Perspectives*. Weinhein, Wiley.

Kar Gupta, A. 2006b. Money exchange model and a general outlook. *Physica A*, 359, 634–640.

Keen, S. 1995. Finance and economic breakdown: modeling Minky's 'financial instability hypothesis'. *Journal of Post Keynesian Economics*, 17, 607–635.

Keen, S. 2000. The nonlinear economics of debt deflation. In Barnett, W.A., Chiarella, C., Keen, S., Marks, R., and Schnabl, H. (eds.), *Commerce, Complexity and Evolution*. Cambridge, Cambridge University Press.

Keynes, J.M. 1937. The general theory of employment. *The Quarterly Journal of Economics*, 51, 209–223.

Klass, O.S., Biham, O., Levy, M., Malcai, O., and Solomon, S. 2007. The Forbes 400, the Pareto power-law and efficient markets. *The European Physical Journal B*, 55, 143–147.

Kleiber, C., and Kotz, S. 2003. *Statistical Size Distributions in Economics and Acturial Sciences*. Hoboken, NJ, John Wiley & Sons.

Krapivsky, P.L., and Ben-Naim, E. 2002. Nontrivial velocity distributions in inelastic gases. *Journal of Physics A*, 35, L147.

Kuznets, S. 1955. Economic growth and income inequality. *The American Economic Review*, 45, 1–28.

Kuznets, S. 1965. *Economic Growth and Structural Change*. New York, Norton.

Lallouache, M., Chakraborti, A.S., Chakraborti, A., and Chakrabarti, B.K. 2010a. Opinion formation in the kinetic exchange models: spontaneous symmetry breaking transition. *Physical Review E*, 82, 056112.

Lallouache, M., Jedidi, A., and Chakraborti, A. 2010b. Wealth distribution: to be or not to be a gamma? *Science and Culture (Kolkata, India)*, 76, 478.

Levy, M. 2003. Are rich people smarter? *Journal of Economic Theory*, 110, 42–64.

Levy, M., and Levy, H. 2003. Investment talent and the Pareto wealth distribution: theoretical and experimental analysis. *The Review of Economics and Statistics*, 85, 709–725.

Levy, M., and Solomon, S. 1997. New evidence for the power law distribution of wealth. *Physica A*, 242, 90–94.

Lorenz, M.O. 1905. Methods of measuring the concentration of wealth. *Publications of the American Statistical Association*, 9, 209–219.

Lübeck, S. 2004. Universal scaling behavior of non-equilibrium phase transitions. *International Journal of Modern Physics B*, 18, 3977–4118.

Lux, T. 2005. Emergent statistical wealth distributions in simple monetary exchange models: a critical review. In Chatterjee, A., Yarlagadda, S., and Chakrabarti, B.K.

(eds.), *Econophysics of Wealth Distributions*. New Economic Windows Series. Milan, Springer Verlag.

Lux, T., and Westerhoff, F. 2009. Economics crisis. *Nature Physics*, 5, 2.

Lydall, H.F. 1959. The distribution of employment incomes. *Econometrica*, 27, 110–115.

Lydall, H. 1968. *The Structure of Earnings*. Oxford, Clarendon Press.

Majumdar, S.N., Krishnamurthy, S., and Barma, M. 2000. Nonequilibrium phase transition in a model of diffusion, aggregation and fragmentation. *Journal of Statistical Physics*, 99, 1.

Malcai, O., Biham, O., Richmond, P., and Solomon, S. 2002. Theoretical analysis and simulations of the generalized Lotka-Volterra model. *Physical Review E*, 66, 031102.

Mandelbrot, B.B. 1960. The Pareto-Levy law and the distribution of income. *International Economic Review*, 1, 79–106.

Marshall, A. 1920. *Principles of Economics*. London; reprinted by Prometheus Books, Macmillan.

Marsili, M., Maslov, S., and Zhang, Y.-C. 1998. Dynamical optimization theory of a diversified portfolio. *Physica A*, 253, 403–418.

Mas-Colell, A., Whinston, M.D., and Green, J.R. 1995. *Microeconomic Theory*. New York, Oxford University Press.

Matthes, D., and Toscani, G. 2008a. Analysis of a model for wealth redistribution. *Kinetic and Related Models*, 1, 1–22.

Matthes, D., and Toscani, G. 2008b. On steady distributions of kinetic models of conservative economies. *Journal of Statistical Physics*, 130, 1087–1117.

McDonald, J. 1984. Some generalized functions for the size distribution of income. *Econometrica*, 52, 647–663.

Meade, J.E. 1964. *Efficiency, Equality and Ownership of Property*. London, Allen and Unwin.

Meade, J.E. 1975. *The Just Economy*. London, Allen and Unwin.

Menchik, P.L., and Martin, D. 1983. Income distribution, lifetime savings, and bequests. *American Economic Review*, 73, 672–690.

Merz, J., Hirschel, D., and Zwick, M. 2005. *Struktur und Verteilung hoher Einkommen. Mikroanalysen auf der Basis der Einkommensteuerstatistik*. Bundesministerium für Gesundheit und Soziale Sicherung.

Mitzenmacher, M. 2004. A brief history of generative models for Power law and lognormal distributions. *Internet Mathematics*, 1, 226–251.

Mizuno, T. 2008. Power law of customers' expenditures in convenience stores. *Journal of Physical Society Japan*, 77, 035001.

Mizuno, T., Toriyama, M., Terano, T., and Takayasu, M. 2008. Pareto law of the expenditure of a person in convenience stores. *Physica A*, 387, 3931–3935.

Modigliani, F., and Brumberg, R. 1954. Utility analysis and the consumption function: an interpretation of cross-section data. In Kurihara, K.K. (ed.), *Post-Keynesian Economics*. New Brunswick, NJ, Rutgers University Press.

Mohanty, P.K. 2006. Generic features of the wealth distribution in ideal-gas-like markets. *Physical Review E*, 74, 011117.

Montroll, E.W., and Shlesinger, M.F. 1982. On 1/f noise and other distributions with long tails. *Proceedings of the National Academy of Sciences USA*, 79, 3380–3383.

Moss de Oliveira, S., de Oliveira, P.M.C., and Stauffer, D. 1999. *Evolution, Money, War and Computers: Non-Traditional Applications of Computational Statistical Physics*. Leipzig, B.G. Tuebner.

Newman, M. 2005. Power laws, Pareto distributions and Zipf's law. *Contemporary Physics*, 46, 323–351.

Nirei, M., and Souma, W. 2007. A two factor model of income distribution dynamics. *Review of Income and Wealth*, 53, 440–459.

Pareschi, L., and Toscani, G. 2006. Self-similarity and power-like tails in nonconservative kinetic models. *Journal of Statistical Physics*, 124, 747–779.

Pareto, V. 1897. *Cours d'economie politique*. Lausanne, Rouge.

Patriarca, M., Chakraborti, A., and Kaski, K. 2004. Statistical model with a standard gamma distribution. *Physical Review E*, 70, 016104.

Patriarca, M., Chakraborti, A., Kaski, K., and Germano, G. 2005. Kinetic theory models for distribution of wealth: power law from overlap of exponentials. In Chatterjee, A., Yarlagadda, S., and Chakrabarti, B.K. (eds.), *Econophysics of Wealth Distributions*. New Economic Windows Series. Milan, Springer.

Patriarca, M., Chakraborti, A., and Germano, G. 2006. Influence of saving propensity on the power law tail of wealth distribution. *Physica A*, 369, 723.

Patriarca, M., Chakraborti, A., Heinsalu, E., and Germano, G. 2007. Relaxation in statistical many-agent economy models. *European Physical Journal B*, 57, 219.

Patriarca, M., Heinsalu, E., and Chakraborti, A. 2010. Basic kinetic wealth-exchange models: common features and open problems. *The European Physical Journal B*, 73, 145–153.

Pianegonda, S., Iglesias, J.R., Abramson, G., and Vega, J.L. 2003. Wealth redistribution with conservative exchanges. *Physica A*, 322, 667–675.

Piketty, T., and Saez, E. 2003. Income inequality in the United States, 1913–1998. *The Quarterly Journal of Economics*, 118, 1–39.

Plischke, M., and Bergersen, B. 2006. *Equilibrium Statistical Physics*. Singapore, World Scientific.

Quadrini, V. 2000. Entrepreneurship, saving, and social mobility. *Review of Economic Dynamics*, 3, 1–40.

Rawlings, P.K., Reguera, D., and Reiss, H. 2004. Entropic basis of the Pareto law. *Physica A*, 343, 643–652.

Rawls, J. 1971. *A Theory of Justice*. Cambridge, MA, Harvard University Press.

Reed, W.J. 2001. The Pareto, Zipf and other power laws. *Economics Letters*, 74, 15–19.

Repetowicz, P., Hutzler, S., and Richmond, P. 2005. Dynamics of money and income distributions. *Physica A*, 356, 641–654.

Rezai, A., Foley, D.K., and Taylor, L. 2012. Global warming and economic externalities. *Economic Theory*, 48, 329–351.

Richmond, P., and Solomon, S. 2001. Power laws are Boltzmann laws in disguise. *International Journal of Modern Physics C*, 12, 333–343.

Richmond, P., Hutzler, S., Coelho, R., and Repetowicz, P. 2006. A review of empirical studies and models of income distributions in society. In Chakrabarti, B.K., Chakraborti A., and Chatterjee, A. (eds.), *Econophysics and Sociophysics: Trends and Perspectives*. Weinheim, Wiley-VCH.

Ross, S.M. 1970. *Applied Probability Models with Optimization Applications*. New York, Dover.

Rutherford, R.S.G. 1955. Income distribution: a new model. *Econometrica*, 23, 277–294.

Salem, A.B.Z., and Mount, T.D. 1974. A convenient descriptive model of income distribution: the gamma density. *Econometrica*, 42, 1115–1127.

Samuelson, P.A. 1998. *Economics*. Auckland, McGraw-Hill.

Scafetta, N., Picozzi, S., and West, B.J. 2004. A trade-investment model for distribution of wealth. *Physica D*, 193, 338–352.

Scruton, R. 1985. *Thinkers of the New Left*. London, Longman.

Sen, A. 1999. *Poverty and Famines*. New Delhi, Oxford University Press.

Sen, P. 2010. Phase transitions in a two parameter model of opinion dynamics with random kinetic exchanges. *Physical Review E*, 83, 016108.

Shorrocks, A.F. 1975. On stochastic models of size distributions. *Review of Economic Studies*, 42, 631–641.

Shorrocks, A.F. 1988. Aggregation issues in inequality measurement. In Eichorn W. (ed.), *Measurement in Economics*. New York, NY, Physica-Verlag.

Shostak, F. 2000. The mystery of the money supply definition. *Quarterly Journal of Austian Economics*, 3, 69–76.

Silva, A.C., and Yakovenko, V.M. 2005. Temporal evolution of the 'thermal' and 'superthermal' income classes in the USA during 1983–2001. *Europhysics Letters*, 69, 304–310.

Silver, J., Slud, E., and Takamoto, K. 2002. Statistical equilibrium wealth distributions in an exchange economy with stochastic preferences. *Journal of Economic Theory*, 106, 417–435.

Simon, H.A. 1955. On a class of skew distribution functions. *Biometrika*, 42, 425–440.

Singh, S.K., and Maddala, G.S. 1976. A function for size distribution of incomes. *Econometrica*, 44, 963–970.

Sinha, S. 2003. Stochastic maps, wealth distribution in random asset exchange models and the marginal utility of relative wealth. *Physica Scripta*, T106, 59–64.

Sinha, S. 2005. The rich are different! Pareto law from asymmetric interactions in asset exchange models. In Chatterjee, A., Yarlagadda, S., and Chakrabarti, B.K. (eds.), *Econophysics of Wealth Distributions*. New Economic Windows Series. Milan, Springer Verlag.

Sinha, S. 2006. Evidence for power-law tail of the wealth distribution in India. *Physica A*, 359, 555–562.

Sinha, S., Chatterjee, A., Chakraborti, A., and Chakrabarti, B.K. 2010. *Econophysics: An Introduction*. Weinheim, Wiley-VCH.

Slanina, F. 2004. Inelastically scattering particles and wealth distribution in an open economy. *Physical Review E*, 69, 046102.

Smith, A. 1776. *An Inquiry into the Nature and Causes of the Wealth of Nations*. London, W. Strahan and T. Cadell.

Solomon, S., and Richmond, P. 2001. Power laws of wealth, market order volumes and market returns. *Physica A*, 299, 188–197.

Solomon, S., and Richmond, P. 2002. Stable power laws in variable economies: Lotka-Volterra implies Pareto-Zipf. *European Physical Journal B*, 27, 257–261.

Sornette, D. 2004. *Why Stock Markets Crash: Critical Events in Complex Financial Systems*. Princeton, NJ, Princeton University Press.

Souma, W. 2001. Universal structure of the personal income distribution. *Fractals*, 9, 463–470.

Souma, W. 2002. Physics of personal income. In Takayasu, H. (ed.), *Empirical Science of Financial Fluctuations: the Advent of Econophysics*. Tokyo, Springer.

Souma, W., and Nirei, M. 2005. Empirical study and model of personal income. In Chatterjee, A., Yarlagadda, S., and Chakrabarti, B.K. (eds.), *Econophysics of Wealth Distributions*. New Economic Windows Series. Milan, Springer Verlag.

Steindl, J. 1972. The distribution of wealth after a model of Wold and Whittle. *The Review of Economic Studies*, 39, 263–279.

Strotz, R.H. 1956. Myopia and inconsistency in dynamic utility maximization. *Review of Economic Studies*, 23, 165–180.

Strudler, M., Petska, T., and Petska, R. 2003. An analysis of the distribution of individual income and taxes, 1979–2001. In *2003 Proceedings of the American Statistical Association, Social Statistics Section*. Philadelphia, PA, American Statistical Association.

Thurow, L.C. 1970. Analyzing the American income distribution. *American Economic Review*, 60, 261–269.

Toscani, G. 2010. Boltzmann legacy and wealth distribution. *Science and Culture*, 76(9–10), 418–423.

Toscani, G., and Brugna, C. 2010. Wealth redistribution in Boltzmann-like models of conservative economies. In Basu, B., Chakravarty, S.R., Chakrabarti, B.K., and Gangopadhyay, K. (eds.), *Econophysics and Economics of Games, Social Choices and Quantitative Techniques*. New Economic Windows Series. Milan, Springer.

United Nations Development Programme. 2004. *Human Development Report*. See http://hdr.undp.org/en/reports/global/hdr2004.

Vaughan, R.N. 1979. Class behaviour and the distribution of wealth. *Review of Economic Studies*, 46, 447–465.

Venkatasubramanian, V. 2010. Fairness is an emergent self-organized property of the free market for labor. *Entropy*, 12, 1514–1531.

Walras, L. 1874–1877. *Eléments d'économie politique pure*. Lausanne, Corbaz.

Wang, N. 2007. An equilibrium model of wealth distribution. *Journal of Monetary Economics*, 54, 1882–1904.

Wang, Y., and Ding, N. 2005. Dynamic process of money transfer models. In Chatterjee, A., Yarlagadda, S., and Chakrabarti, B.K. (eds.), *Econophysics of Wealth Distributions*. New Economic Windows Series. Milan, Springer.

Wang, Y., Xi, N., and Ding, N. 2006. The contribution of money-transfer models to economics. In Chakrabarti, B.K., Chakraborti, A., and Chatterjee, A. (eds.), *Econophysics and Sociophysics: Trends and Perspectives*. Weinheim, Wiley.

Wasserman, S., and Faust, K. 1994. *Social Network Analysis*. Cambridge, Cambridge University Press.

Wilhelm, M.O. 1996. Bequest behavior and the effect of heirs' earnings: testing the altruistic model of bequests. *American Economic Review*, 86, 874–892.

Wold, H.O.A., and Whittle, P. 1957. A model explaining the Pareto distribution of wealth. *Econometrica*, 25, 591–595.

Xi, N., Ding, N., and Wang, Y. 2005. How required reserve ratio affects distribution and velocity of money. *Physica A*, 357, 543–555.

Yaari, M.E. 1965. Uncertain lifetime, life insurance, and the theory of the consumer. *The Review of Economic Studies*, 32, 137–150.

Yakovenko, V.M. 2012. Applications of statistical mechanics to economics: entropic origin of the probability distributions of money, income, and energy consumption. See http://arxiv.org/abs/1204.6483v1.

Yakovenko, V.M., and Barkley Rosser, J., Jr. 2009. Colloquium: statistical mechanics of money, wealth and income. *Reviews of Modern Physics*, 81, 1703–1725.

Yakovenko, V.M., and Silva, A.C. 2005. Two-class structure of income distribution in the USA: exponential bulk and power-law tail. In Chatterjee, A., Yarlagadda, S., and Chakrabarti, B.K. (eds.), *Econophysics of Wealth Distributions*. New Economic Windows Series. Milan, Springer Verlag.

Zhu, J. 2010. Wealth distribution under idiosyncratic investment risk. New York University, Technical Report.

Ziff, R.M. 1984. Aggregation kinetics via Smoluchowski's equation. In Family, F., and Landau, D.P. (eds.), *Kinetics of Aggregation and Gelation*. Amsterdam, North Holland.

Index

For EU product safety concerns, contact us at Calle de José Abascal, 56–1°,
28003 Madrid, Spain or eugpsr@cambridge.org.

www.ingramcontent.com/pod-product-compliance
Lightning Source LLC
LaVergne TN
LVHW062209200926
848484LV00006B/1362
* 9 7 8 1 1 0 7 0 1 3 4 4 5 *